I0822943

IN FINE DETAIL

PERSPECTIVE

For Artists, Architects, and Creatives

Simone Ridyard

DAVID & CHARLES
— PUBLISHING —

www.davidandcharles.com

CONTENTS

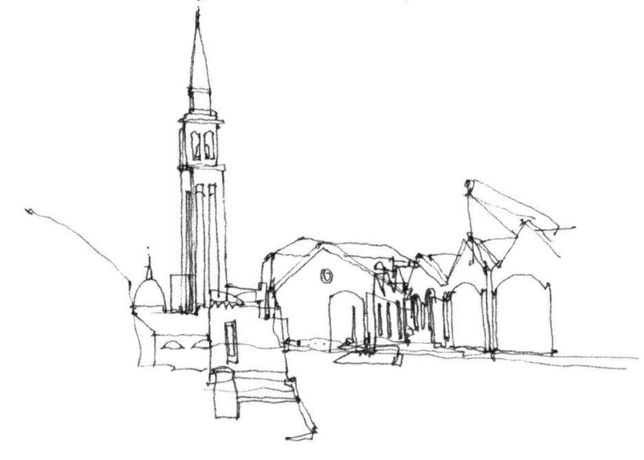

INTRODUCTION

I studied architecture at Canterbury College of Art, as it was known then (now part of the University for Creative Arts). Wednesdays were spent sketching on the streets of the historic cathedral city, followed by life-drawing classes back on campus in the late afternoon. I feel enormously privileged to have had that kind of art-school education, where both the understanding and the love of drawing in perspective were ingrained at an early age. It is an enthusiasm I have never lost.

During my years in architectural practice, I was often involved in the early concept and visualization stages of projects, so drawing remained central to my work. However, it was my later move into academia, and my discovery of the remarkable global Urban Sketchers community, that brought me back to drawing on location. Drawing for pleasure, rather than for work, brings a different kind of satisfaction – and, I would argue, a deeper connection to the act of seeing.

Alongside this, I have greatly enjoyed teaching drawing workshops, where I continually return to the challenge of simplifying the principles of perspective. Any landscape or urban sketch becomes more convincing when it follows the basic rules of linear perspective – a way of representing space so that objects appear smaller as they recede, creating a sense of depth and realism.

This book is an opportunity to explore those principles more deeply. I am delighted to include contributions from a number of highly respected artists and specialists in perspective drawing, whose expertise and generosity have added so much to this book.

For students of architecture and design, it is of course possible to generate perspective views using digital software. But in doing so, you risk missing something important: the sense of achievement – and yes, the joy – that comes from understanding and constructing perspective yourself. Some of the most exciting student work I have seen in recent years combines the immediacy and sensitivity of freehand drawing with the precision of digital programs such as Adobe Photoshop and Procreate. This hybrid approach offers a powerful and relevant way forward, and one that clients consistently value.

The book begins by introducing the fundamental ideas needed to understand how perspective works, before looking in detail at one-point and then two-point perspective. From there, we explore a wider range of perspective situations and their variations. As the book progresses, I introduce the use of diagonals and work through increasingly complex examples, helping you build confidence and develop a deeper understanding of perspective drawing.

Throughout the book, photographs and diagrams include overlays to help explain the underlying structure. The eye-level (or horizon) line is shown as a solid red line. Lighter red lines are usually used to indicate recession towards vanishing points, although in some cases I have used bolder dashed lines to make the construction clearer.

I would like to thank everyone who has contributed to this book, especially Phil Griffin for his wonderful prologue to this fascinating subject, and all those who have supported and encouraged my drawing over the years. In particular, I owe a great deal to the Urban Sketchers community. Without that inspiring network of generous and like-minded artists, this book would not exist.

Concept sketch for Edinboro Castle, Camden, London. Student project completed during my time at Canterbury School of Art.

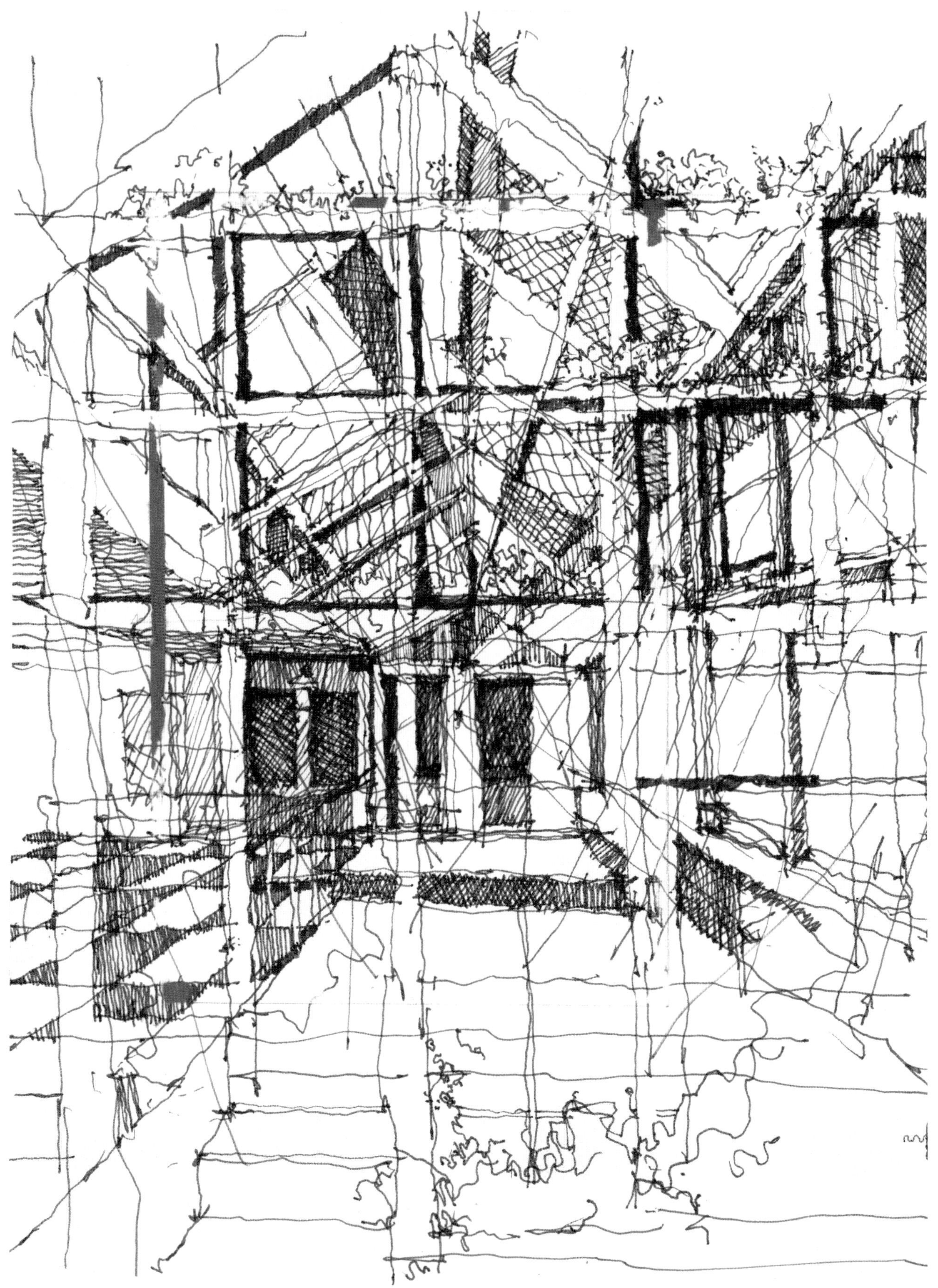

PROLOGUE

"We only see what we look at. To look is an act of choice."

John Berger, Ways of Seeing, 1972

Artists aim to communicate the world they are looking at. Most will aim to do this accurately, truthfully, and attractively. The use of perspective techniques will help to do this. For architects and designers, perspective techniques are essential in expressing ideas and solutions, and having them realized or built out in three dimensions.

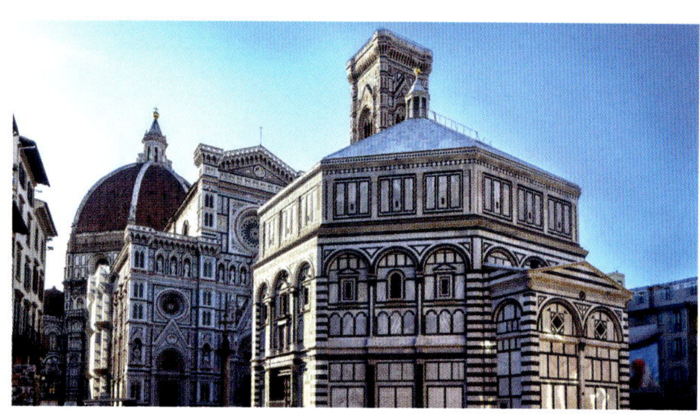

Brunelleschi, an architect and engineer, won the competition to design and build the self-supporting dome of Florence Cathedral and, during this time, conducted a pioneering experiment in linear perspective through a drawing of the Baptistry.

ORIGINS OF LINEAR PERSPECTIVE

You might, if you wanted to make a bold assertion, say that linear perspective – our adopted visual language for expressing and depicting the world, the "lingua franca" of Western art – was coined, codified, and demonstrated by Filippo Brunelleschi (1377–1446) in 1420, when he painted two panels with views of Florence. Like the Renaissance itself, intervening centuries of scholarship and debate have interrogated this assertion, but a large sector of the world will, at the very least, get the point: linear perspective began in Florence in 1420, and it was invented by Filippo Brunelleschi.

Leon Battista Alberti, born in Genoa in 1404, was a Renaissance man, of the full spectrum. He was a polymath whose wide range of interests is hard to grasp – difficult to put in perspective, so to speak. He was a notable architect, poet, linguist, sculptor, painter, and cryptographer. He was also ordained into the Church. A priest who built churches in Florence, Rimini, and Mantua. As a priest, he was a notable humanist. His book, *Della Pittura* [On Painting] (1435), emphasizes his belief in mathematics as the foundation of all science and arts. As such, he studied that most fundamental of Renaissance sciences, optics. His book is often taken to be the first on painting theory. Alberti dedicates *Della Pittura* to Brunelleschi. He writes that "painting is an imitation of reality".

In the states of Italy in the fifteenth century, great works of art and architecture from the period rose to command the story of visual expression, perhaps fuelled by mercantile wealth combined with religious rigour and stability. Northern European artists travelled to Italy to learn from what they saw. So much was happening throughout the fifteenth and sixteenth centuries, but paintings, sculpture, and the great buildings in which they are exhibited do not travel. They are not reproduced in photographs, printed, and distributed throughout the world. But they are transcribed, sketched, and annotated in the journals and notebooks of interested travellers.

Rome began to recover its ancient status and command of (the yet to be unified nation of) Italy. The Catholic Church and the Papacy had been in an uneasy relationship with the Holy Roman Empire for a millennium. The Church, however, continued as a commissioner, enabler, and patron of arts and sciences, if not always to the greater good of the artists and scientists it commanded.

Three masters competed with each other, though indirectly, and not all in Rome at the same time. Leonardo da Vinci (1452-1519) was in Rome briefly in the early 1500s, visiting from Florence. Michelangelo (1475–1564) arrived in Rome in 1508 to begin work on his commission to paint the Sistine Chapel. Raphael (1483–1520) was working in the papal apartments at the same time.

Raphael's fresco *The School of Athens* (1509–1511), painted for Pope Julius II's private library in the Apostolic Palace (in what is now the Vatican City), is a striking example of one-point perspective organizing both the space it depicts, in volume and depth, and the hierarchy of the society and cast it assembles. The one-point perspective frames and orders the narrative of the painting at the same time. The vanishing point is just below Plato's outstretched right hand. Raphael's handling of perspective allowed him to direct his narrative. This, combined with his use of colour, is the order that began to fix our visual language.

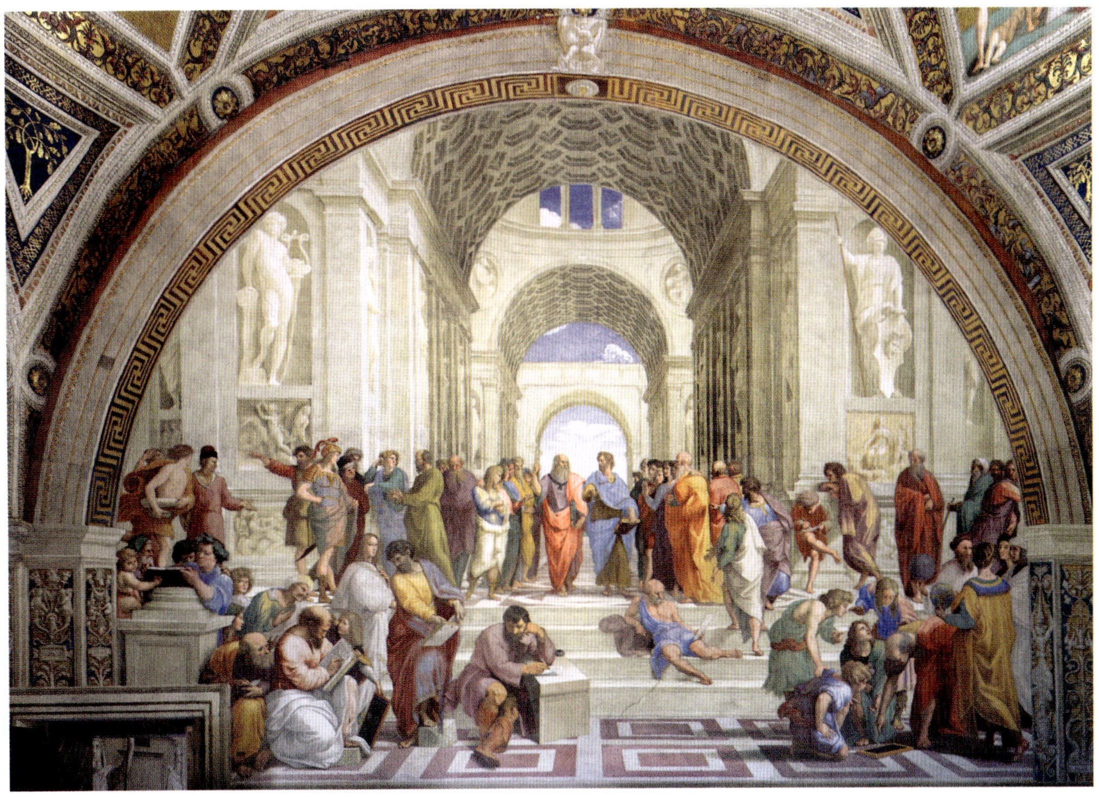

Raphael, The School of Athens, *1509–1511*

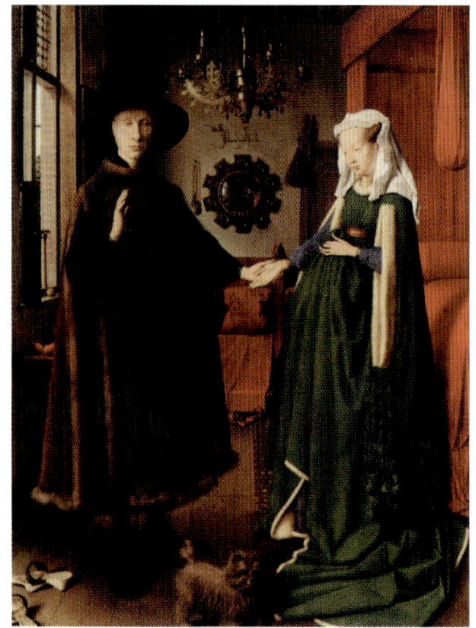

Jan Van Eyck, The Arnolfini Portrait, *1434*

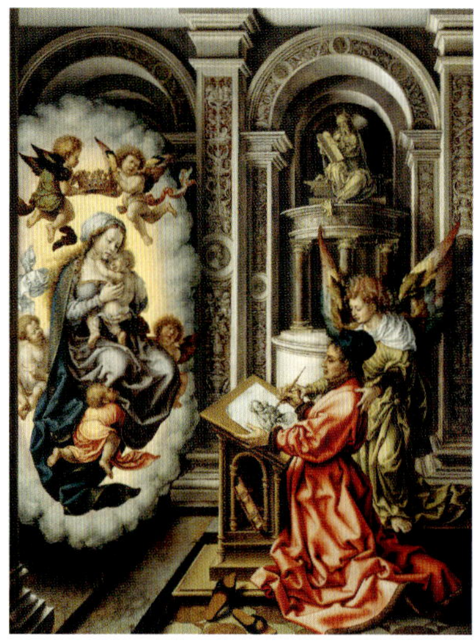

Jan Gossaert, St Luke Painting the Virgin, *c. 1520*

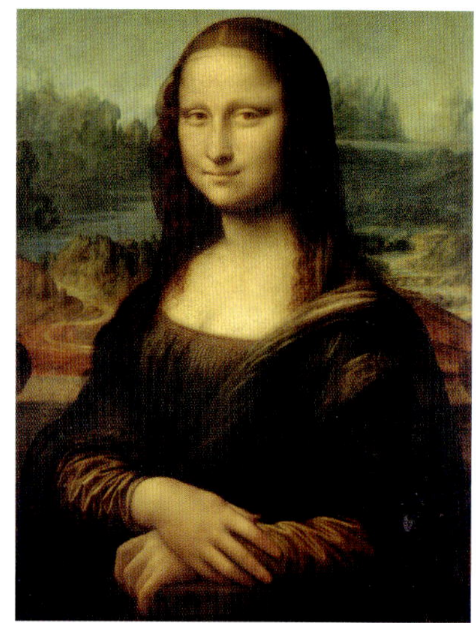

Leonardo da Vinci, Mona Lisa, *1503–1517*

Flemish artists of the period were more empirical. There might have been more depth to their thinking than there was in their natural environment. Jan Van Eyck (1390–1441), in *The Arnolfini Portrait* painted in 1434, ordered a small Dutch interior into a world of symbol and metaphor. Perspective was not his concern, but he used it as a tool to set and frame his (ambiguous and subjective) messages. Perspective is a mechanism for planning a painting – interior or landscape – and rendering it understandable, believable, and pleasing.

In *St Luke Painting the Virgin*, Jan Gossaert (1478–1532) played a great game with perspective, comparing the imaginary heavenly space that the Virgin occupies with the ordered, ecclesiastical surroundings of the saint's reality, rendered in strict perspective.

At a similar time, the painter, architect, engineer, and polymath Leonardo da Vinci sketched a Florentine noblewoman, Lisa del Giocondo. The painted portrait he made, probably completed in 1517, is known as *Mona Lisa*. Arguably, it is the most famous painting in the world, judging by the crowds thronging to see it in the Louvre Museum, Paris. Lisa, with her enigmatic smile, floats before a carefully plotted perspective landscape. The paint itself, its colour, tone, and weight, create what might be called "atmospheric perspective".

Looking at a portrait, particularly one so seemingly direct as *Mona Lisa*, might seem odd when talking about perspective, but it is the carefully considered and constructed background, in such a relatively small picture, that allows Lisa del Giocondo to come forward, almost to escape the frame. In cinema language, *Mona Lisa* is a medium close-up. The background, which da Vinci invented, then constructed and painted, is what allows the subject to come forward, towards the viewer, in such a "present" way. She enters the room.

PERSPECTIVE AS DECEPTION

Paintings can be many things, separately, in combination, or all at once. A painter's depiction of the scene they are looking at, is a painted metaphor for reality. There can be signs, symbols, and narratives, disguised, or in full view. When Raphael painted *The School of Athens* he was not in it. It was in him. It is a work of creation of the imagination. Therefore, he was in the business of making things that never happened (in the way they are depicted), happen for the viewer. Perspective becomes part of the tricks of the artist's trade – that of transporting the viewer to a place, atmosphere, and view that did not exist. Mastery of perspective might also allow the artist to improve on the view.

Giovanni Antonio Canal (1697–1768), also known as Canaletto, painted in a number of places, including Rome and London, but is best known as a painter of his home city of Venice. His paintings are both real and imaginary. Canaletto's mastery of perspective allows him to improve on any clumsy, discordant moments in reality, and reorder his idealized view, moving buildings and gondolas, bringing structures closer to or further away, making them bigger or smaller, into more pleasing arrangements. Each of the countless millions of visitors to Venice in the last 250 years will have had a preconceived picture of the Grand Canal drawn from some remembered version by Canaletto.

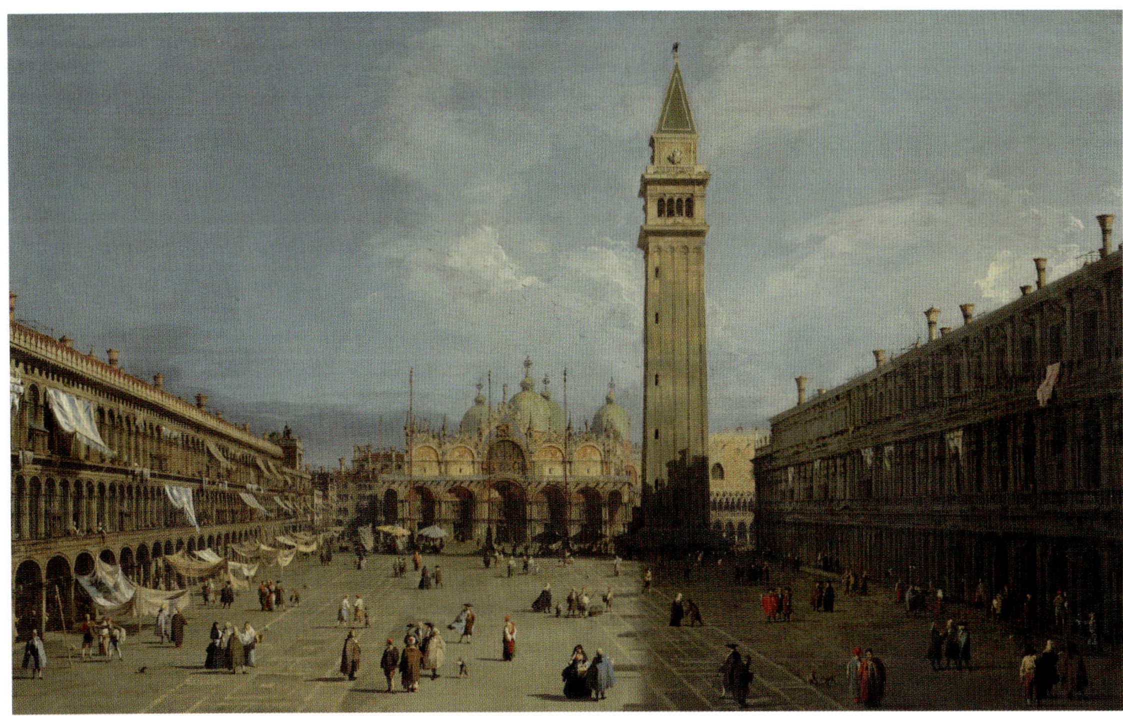

Canaletto, The Piazza San Marco in Venice, *c. late 1720s*

Rules of perspective are there to be broken, as the many abstract movements in painting amply demonstrate. Renaissance artists themselves bent the rules, even as they were forging them. Fresco painters, with their mastery of perspective, made domes out of flat surfaces, opened ceilings to the heavens, and brought vast exterior landscapes into small interior spaces. Knowing the rules is the first step to manipulating them to dramatic effect.

Of the many examples of deceptive perspective, through the many movements in art, none is more bravura than Andrea Pozzo's frescos in Sant'Ignazio, the church in Rome that Jesuits built to honour their founder Ignatius Loyola. The Jesuits were given a huge papal grant when Ignatius was canonized in 1622. They planned a massive church in his (and their own) honour, but had exceeded their budget by 1642, and couldn't afford the planned grand dome. Their loyal lay brother, Andrea Pozzo (1642–1709), was working on his two-volume *Perspectiva pictorum et architectorum* (1693–1700) when he was put in charge of the interior decoration of Sant' Ignazio in 1684. Jesuits got their dome, the whole globe, and the heavens too. All are rendered on the flat interior surfaces of the basilica, using the illusionistic techniques known as *trompe l'oeil* (French for "deceive the eye").

Deception, illusion, and tricking the eye are not primarily the aim in applying the rules of perspective. Though Canaletto occasionally reordered the palazzo and piazzas of Venice, he did so to make his paintings the more pleasing to viewers who may never come within a thousand miles (or a hundred years) of the view. Andrea Pozzo used his techniques to give the Jesuits their dome, and Saint Ignatius his cloud in heaven. Sometimes, perspective technique is downright trickery. A century before Pozzo conjured his dome in Rome, Andrea Palladio (1508–1580) created a vast stage set within a small theatre in Vicenza.

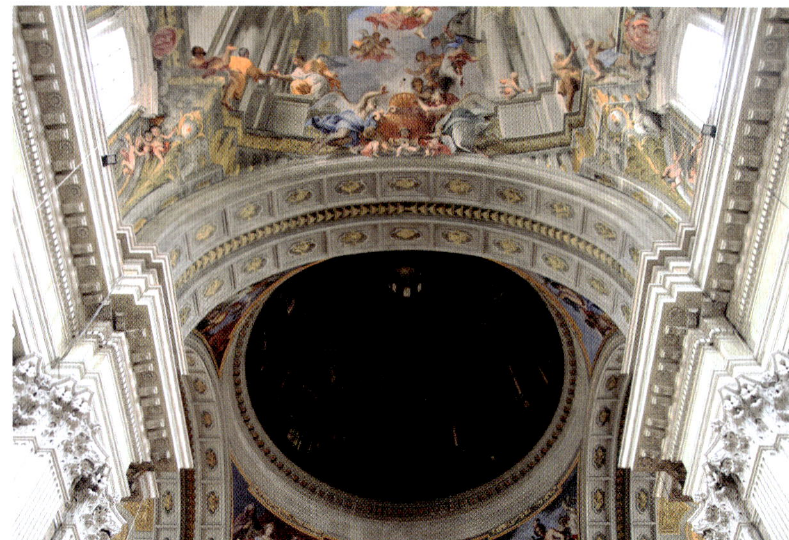

Andrea Pozzo, trompe l'oeil cupola of Sant'Ignazio church, *Rome, Italy, 1685*

Teatro Olimpico in Vicenza is a box of visual tricks – intentionally and flamboyantly so. The permanent on-stage scenery, Roman-style back screen, full-width of the stage, is painted wood and stucco, resembling marble, designed by Vincenzo Scamozzi (1548–1616). It is the oldest surviving stage set in the world, still in use, and the effect takes the audience from elaborate decorated and statued interior, down long streets, stretching to the far horizon. *Trompe l'oeil*, *coup de théâtre*, perspective magic.

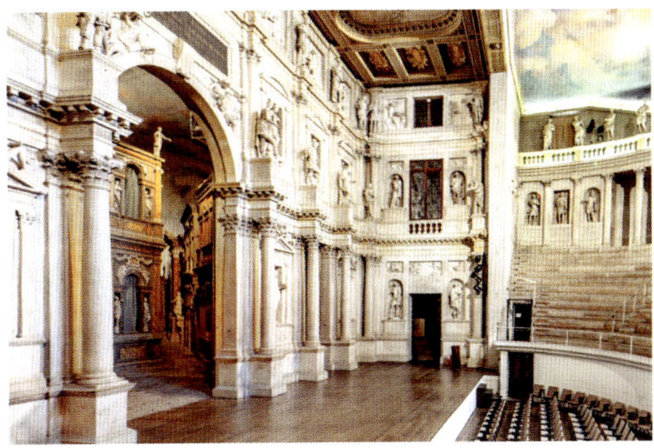

Teatro Olimpico, *Vicenza*, *Italy*

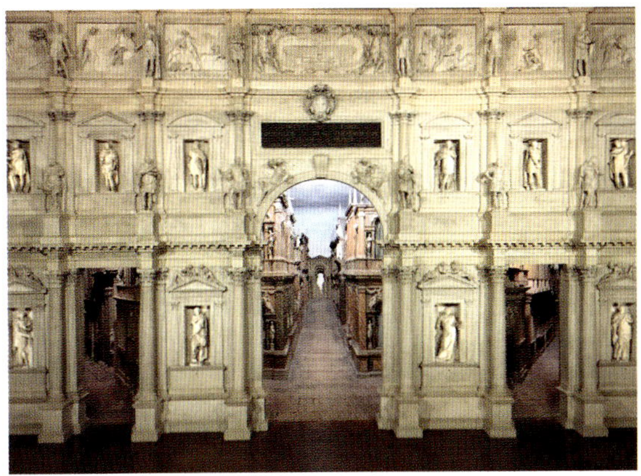

Teatro Olimpico, *Vicenza*, *Italy*

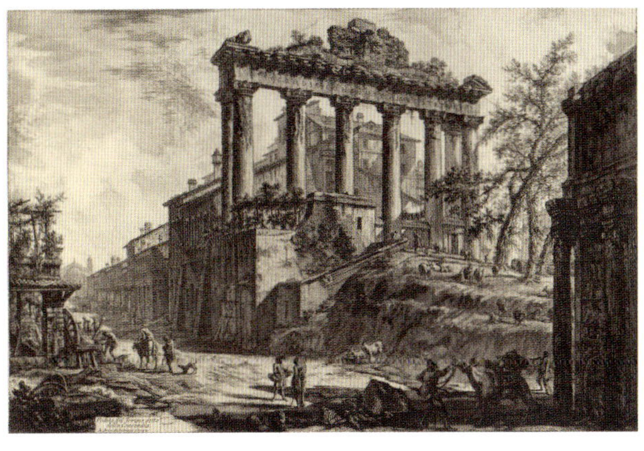

Giovanni Battista Piranesi, View of the Temple of Concord, Rome, c. 1740–1760

There are artists who conjure with perspective, mathematics, optics, and the imagination. Giovanni Battista Piranesi (1720–1778) was an artist, archaeologist, architect, and engraver working in Venice and Rome. His series *Le Vedute di Roma* [Views of Rome] (1745) both recreated and reimagined the ancient city in reproduceable editions. He played with mathematics and perspective to produce highly atmospheric imaginary interiors, proving that formal architectural spaces can hold compelling imaginary narrative. Piranesi said nothing that the early Renaissance artists weren't capable of expressing, with the addition of a vivid, even febrile, imagination. His series of etchings *Carceri d'invenzione* [Imagined Prisons] demonstrated depth and scale in rigorous perspective, with the addition of masterly chiaroscuro. Piranesi has inspired so much in theatre, set design, and cinematography, he might easily be seen as a precursor to film noir.

THE IMPACT OF PHOTOGRAPHY

The rules of perspective help the artist present verisimilitude. They help designers and architects present ideas to fill spaces. They also present opportunities for manipulation, exaggeration, and deception. Joseph Nicéphore Niépce (1765–1833) and Louis Daguerre (1787–1851) developed their daguerreotype process by 1839, improving the quality of the photographic image and making it durable and reproducible. Photography changed everything. Taken in combination with railway travel, the second half of the nineteenth century was monumental. It changed all our perspectives. Literally and figuratively.

Eventually, the photographic image moved. When photography becomes film, and cameras are developed that will take multiple images in sequence, and perforations allow film to be transported through a camera, doubling as projector, the movies are born. The Lumière brothers held their first public screening in Paris in 1895: the sequence "Arrival of a Train". It was footage of a train pulling into Gare de La Ciotat, the station in La Ciotat, near Marseille. Filmmakers have been playing with perspective ever since. Cameras, the railways, and movies opened a world that invited new ways of seeing. Perspective provided the bridge between the wide expanses of director John Ford's Midwest (*The Searchers* and *How the West Was Won*) and the claustrophobia of fictional private investigator Philip Marlowe's mean streets (*The Big Sleep* and *The Long Goodbye*). Movie audiences distinguish between scales and distances because of perspective.

At the same time, the painters whom we bracket as the Impressionists, were dealing with light (often around smoky railway stations in Paris) in layers of paint and techniques that define atmospheric perspective. The Impressionists held the ground for colouration while photography was in its black and white period. Perspective provided the bridge between the tight gridded canyons of the Eastern Seaboard cities and the vast divergent landscapes of the unfolding west of the United States. After the American Civil War, artists of the Hudson River School headed West from New York and painted the emerging continent as though it were the interior of some vast topographical cosmos. Albert Bierstadt (1850–1902), originally from Prussia, painted mountains in a haze of pigments. His atmospheric perspective paintings opened up the Midwest faster than the Union Pacific Railroad.

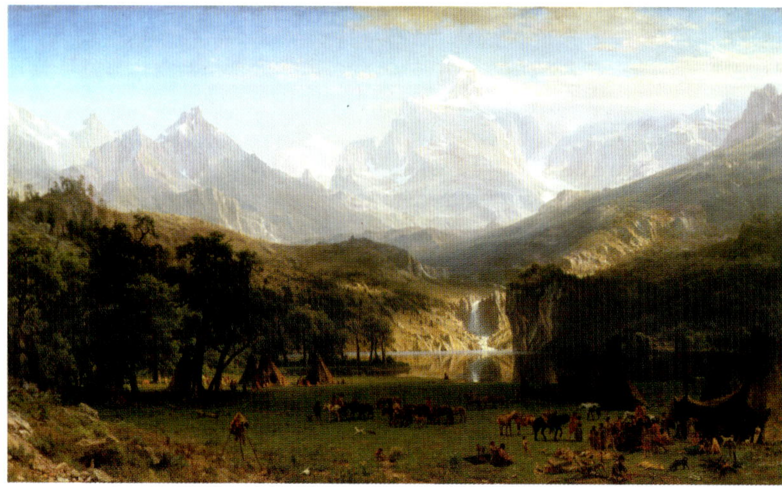

Albert Bierstadt, The Rocky Mountains, Lander's Peak, *1863*

In the industrial mill towns and engineering conurbations of the north of England, the enclosure of streets around factories and football grounds became the now-familiar vista of Laurence Stephen Lowry (1887–1976). Long under-appreciated for the skilled painter and draughtsman he was, Lowry reproduced his deceptively simple street scenes with repetitive precision. Whether it is *Piccadilly Circus* or *Going to the Match*, his linear perspective holds a seemingly simple representational composition, which he then populates with characters of the artist's own scale and preference; buses, prams, dogs, and all. And the artist was able to reproduce the smog and haze of his lived streets through his technical understanding of atmospheric perspective.

Debates between black and white vs colour will be infinite. Meanwhile, both means of expression will endure, with fans on either side of the debate. It's notable that polls of the world's greatest movies, up to the present day, regularly place *Citizen Kane*, written and directed by Orson Welles, at number one. Welles and his cameraman Gregg Toland did something extraordinary. Together, auteur and cinematographer made perspective into a major cinematic creative tool. They foreshortened sets, animated the inanimate, dwarfed and grew their performers at will, and created vast interiors disappearing into infinity.

David Hockney arrived in Los Angeles in 1964. When not painting swimming pools, he was driving the Hollywood Hills and the Sierras enclosing the valley. His pictures took a quintessentially English atmosphere of rolling hills and cultivated fields, dissected by roads and highways. As ever, he played with perspective, making the landscape rise and fall, tumbling like a roller-coaster. Later in his life, he began to question not how a painter can do this, but why this manipulation of perspective came to define Western art. As ever, over a long life, David Hockney has investigated how we look, and how we express, what we see. "To look is an act of choice". First you look, then you let perspective take it from there.

Phil Griffin

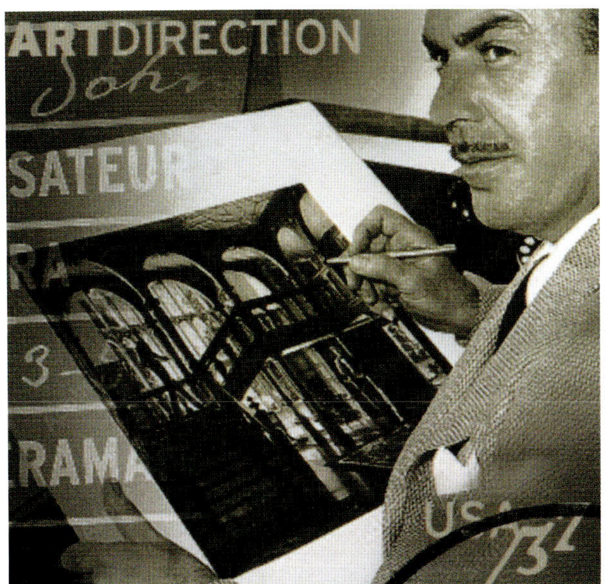

Art director Perry Ferguson working on sketch for Citizen Kane, American Filmmaking Behind the Scenes, 2003

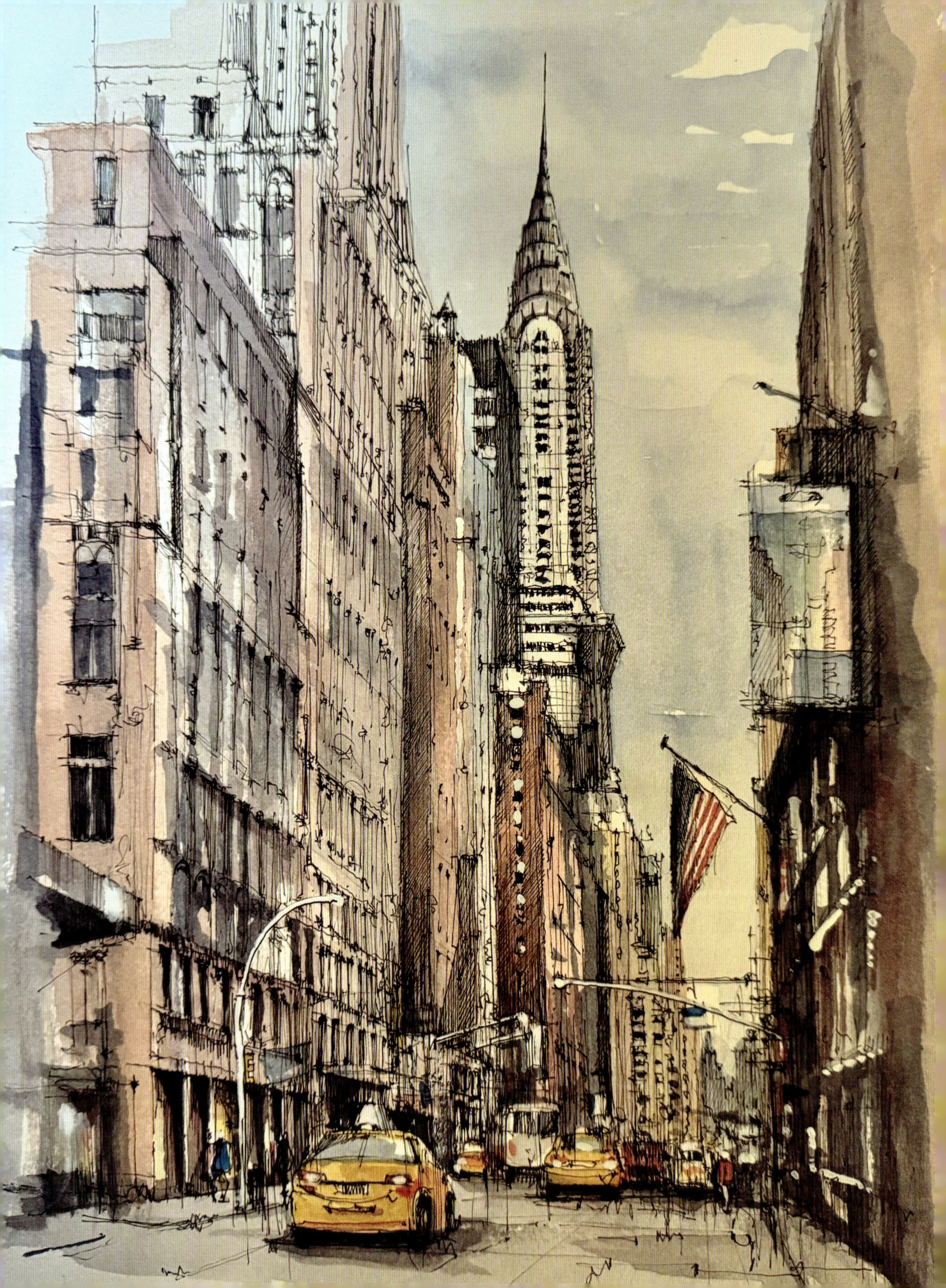

01

RULES OF DRAWING

Many people feel anxious about perspective – even experienced artists. With my training as an architect, I'm very comfortable with its principles. While a deep understanding isn't essential for making art, a basic awareness of the principles is very useful. You don't need to master all the complexities; the aim is simply to use perspective to suggest three-dimensional space on a two-dimensional surface.

When we draw a view, our aim is to recreate on paper or canvas what we see before us. This might be done directly on location or by working from photographs. To achieve this, we rely on visual clues in the scene – such as massing, scale, composition, and depth. We observe what is closest to us and what lies further away, and we think about how the different elements relate to one another in space.

Perspective not only makes a scene feel more realistic but also suggests how objects relate to one another in space, giving a better sense of depth, distance, and placement within the composition. Before engaging with the technical aspects of perspective drawing, it's important to first understand what we actually "see". By establishing some basic principles, we can build a framework that will help us as we move from simple to more complex types of perspective drawing.

A one-point perspective street view of New York City, with the Chrysler Building visible in the distance. Fineliner and watercolour by Caroline Smith.

NATURAL PERSPECTIVE

When we look at a group of objects, such as buildings of the same size, those closer to us appear larger and those further away appear smaller. Even though we know they are equal in size, our eyes see them differently. This effect is called natural perspective. It explains how we naturally experience depth and distance without using formal drawing rules. In art, natural perspective helps create a realistic sense of space and depth.

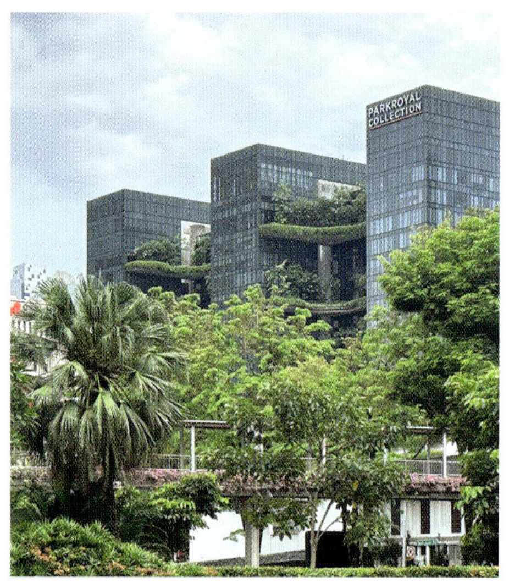

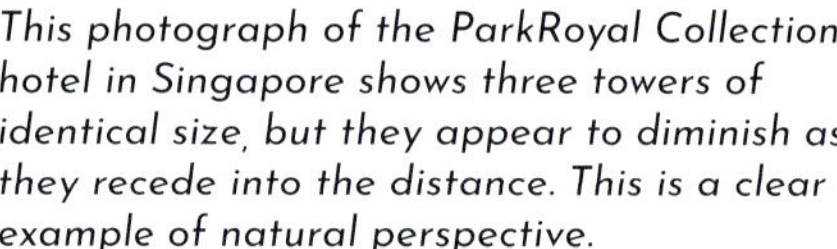

This photograph of the ParkRoyal Collection hotel in Singapore shows three towers of identical size, but they appear to diminish as they recede into the distance. This is a clear example of natural perspective.

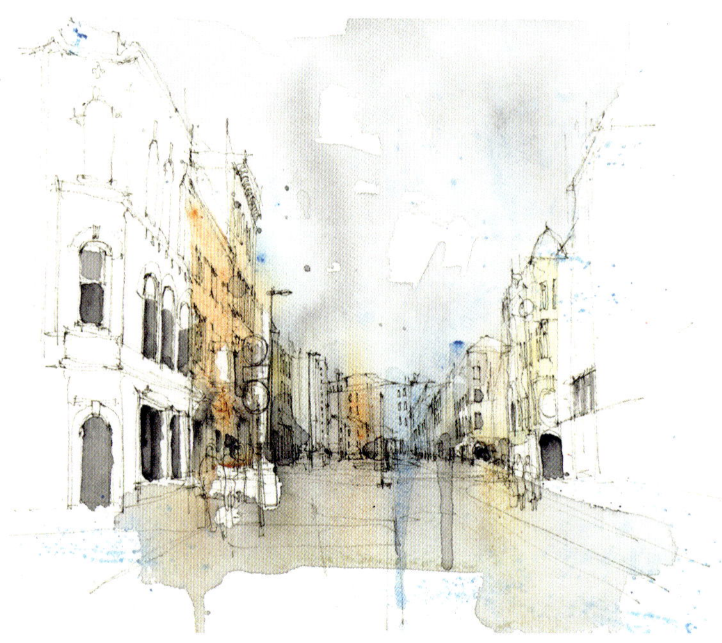

This sketch of Stephenson Square, Manchester, UK, is also an example of natural perspective. Buildings of similar height, mostly three to four storeys, appear smaller as they recede naturally into the distance.

A classic example is lines of telegraph poles along a highway. We know they are all the same height, yet they appear smaller as they stretch into the distance. This is natural perspective – an optical illusion created by depth and distance. In scenes like this, we focus on the point where the objects (for example telegraph poles) appear to meet. This is called the vanishing point, and it sits at eye level (also called the "horizon line"). Understanding this is essential for learning the basics of perspective drawing.

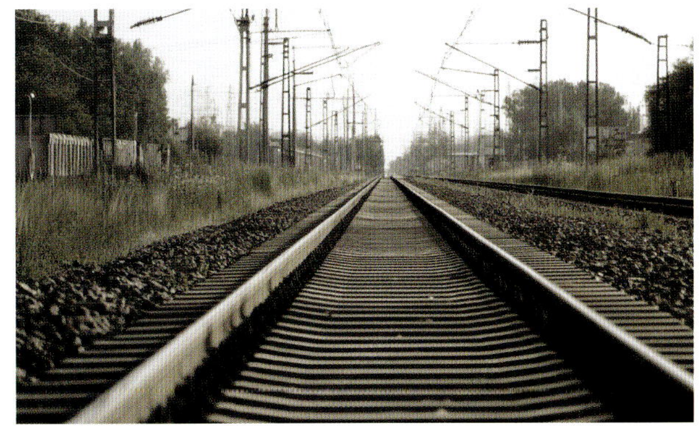

Another clear example is provided by the railway tracks shown here. Although we know they run parallel, they appear to converge at the horizon, demonstrating natural perspective.

GROUND PLANE VS EYE LEVEL

The term ground plane describes a physical surface that forms the "floor" of a scene. While the ground plane is something you might actually draw as part of the image, eye level is more abstract, relating to you and indicating the height of your viewpoint.

The ground plane is clearly visible in this sketch of Washington Park Square, New York. The park itself forms the ground plane, with the lampposts and trees sitting on it, while eye level is positioned above.

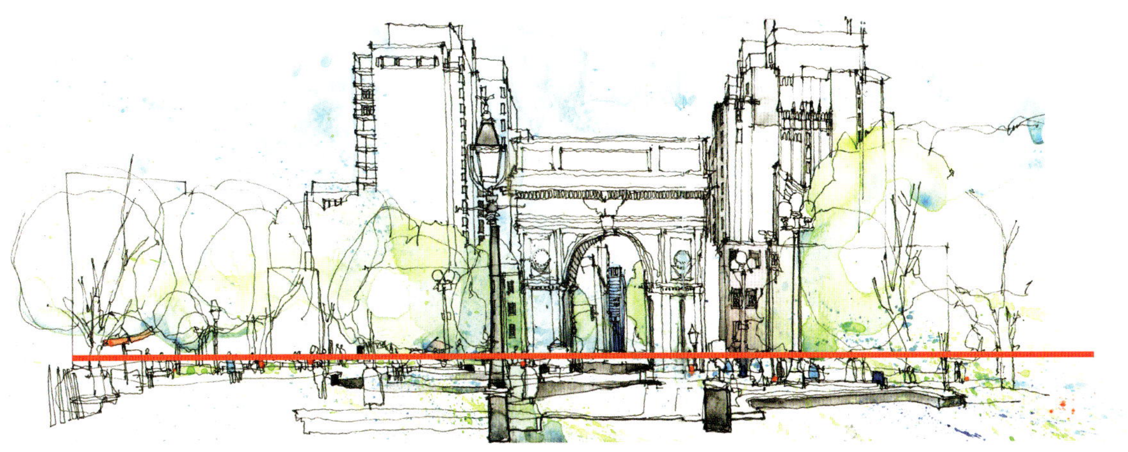

EYE LEVEL

Let's look at eye level in more detail. To start, we need to establish our line of sight – the direction in which we are looking. In one-point perspective (see Chapter 02), this line runs straight ahead, parallel to our position.

CONE OF VISION

Closely related to the eye level is the cone of vision, which defines the area we can see clearly. This "cone" spans between 45 to 60 degrees and marks the limits of our accurate sight-line. While we also have peripheral vision, when applying perspective rules, we focus only on the cone of vision, sometimes called the field of view.

PICTURE PLANE

Assuming we are on a level surface (the ground plane), then we must establish what is known as the picture plane. Think of this like looking through a window. The window glass serves as the picture plane, capturing a three-dimensional view that will be represented in two dimensions using perspective rules. It is important to understand that the picture plane is perpendicular to our cone of vision – simply put, this means it is flat and faces us directly. In practice, the picture plane is the same as the actual surface of your sketch, drawing, or painting. The cone of vision will intersect with the picture plane. It is important to consider eye level at the point where the cone of vision meets the picture plane.

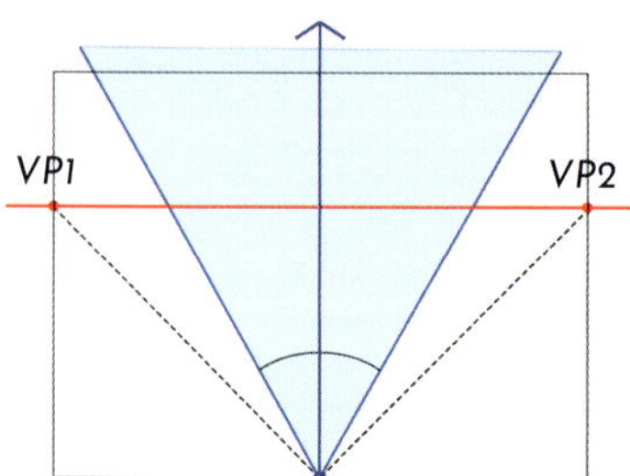

This diagram shows the cone of vision on a flat plane. The blue dot represents the viewer's position (station point), and the triangular cone represents normal vision. The red line represents eye level. Using peripheral vision, two vanishing points (VP1 and VP2) can be seen.

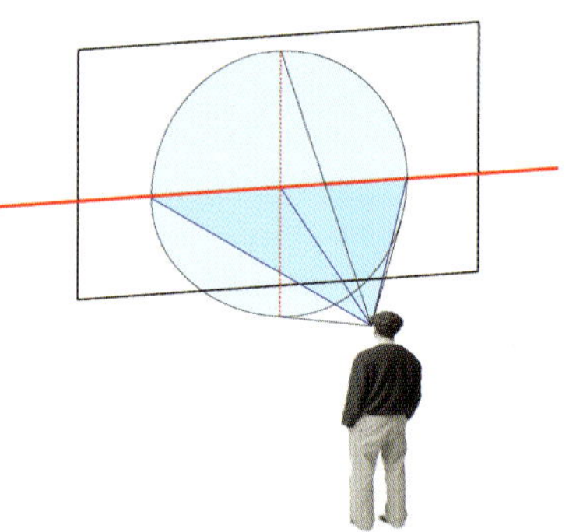

This diagram illustrates the picture plane in three dimensions and shows a viewer positioned on the ground plane at the station point. From this position, the viewer looks towards the picture plane, with their field of vision defined by the cone of vision. The red line represents eye level.

HOW TO FIND YOUR EYE LEVEL

Getting the correct eye level is essential for using perspective properly. First, think about where the horizon is in your view. This shows your eye level, which depends on your position – for example, are you standing or seated?

You can easily find your eye level by holding a pen or pencil perpendicular to your eyes with your arm outstretched and note where it intersects with the horizon. This intersection marks *your* eye level, which you can then draw on your paper. With the eye level established, you can determine what elements in the view are positioned below or above it.

As an approximate guide, when a person is standing, their eye level is usually between 1.5–1.6m (4ft 11in–5ft 2in). When sitting, it drops to 1.2m (3ft 11in) for adults. The key point to remember is simple: your eye level changes with your position.

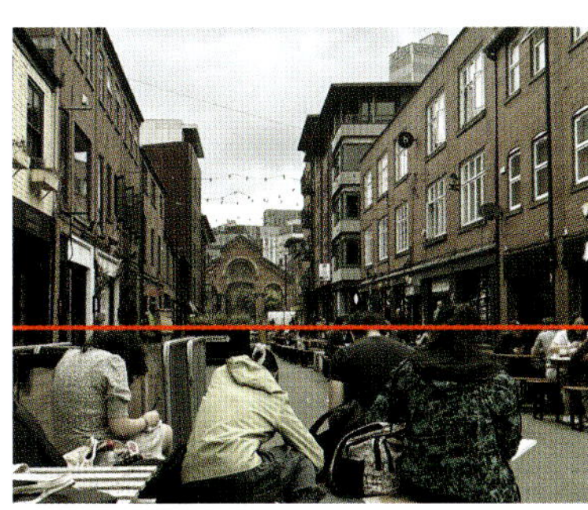

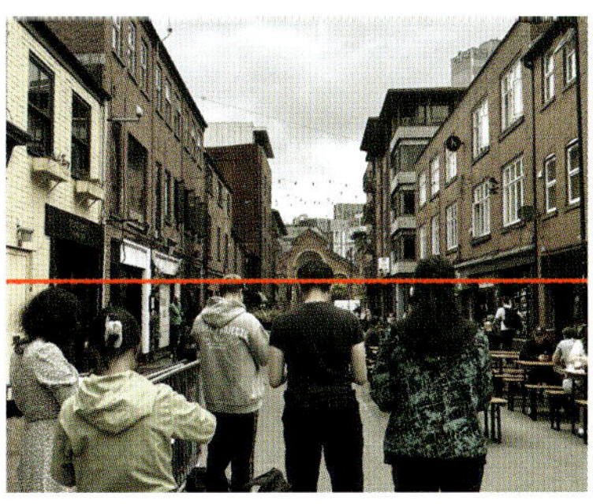

In the first photograph of students sketching, the students are seated, so the eye level is lower. In the second photograph, they are standing, which raises the eye level. It is important to understand that these eye-level lines are indicative, as eye level will be slightly different for everyone.

Because I was sitting down when sketching this view of the Three Graces on the Liverpool Waterfront, UK, my eye level was relatively low. Notice how the buildings diminish in size and detail as they recede into the distance, compared with the sharper detail of those closest to me.

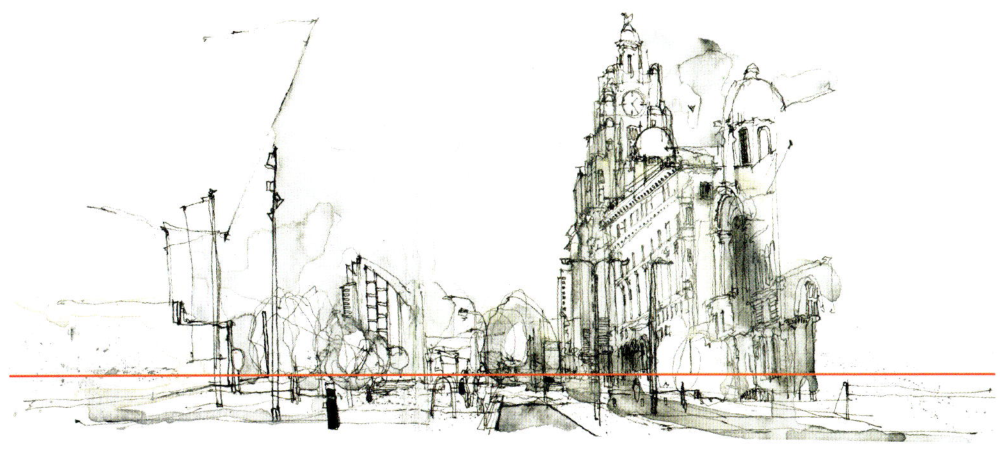

CHANGING THE EYE LEVEL

Building on these basic principles: if you're sketching at second-floor level, the ground plane remains exactly that – the ground – while your eye level adjusts to match the height of the second floor.

You can see this illustrated in a sketch from the second floor of the Museum of Modern Art (MoMA) in New York, where the ground plane is clearly identified as the ground level (shown shaded).

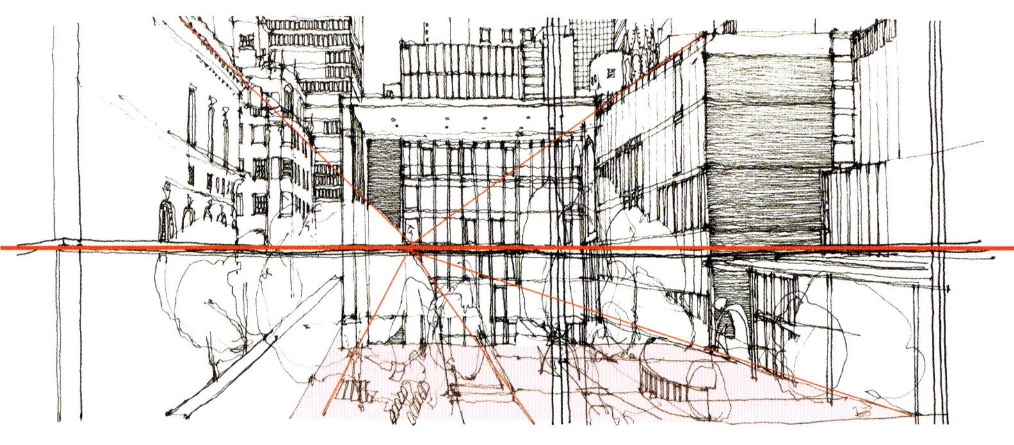

MoMA, New York, USA

When looking down a steep hill, such as in these Lisbon examples, the ground plane is hard to see because it slopes steeply away.

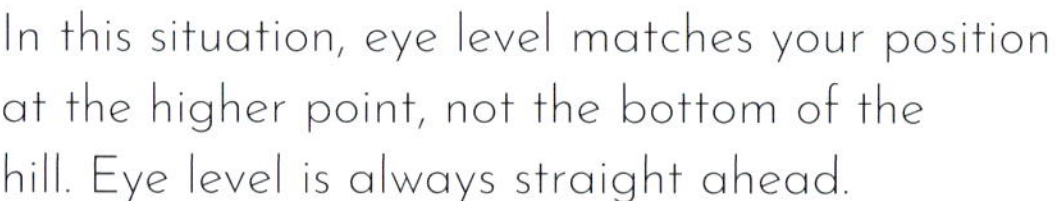

In this situation, eye level matches your position at the higher point, not the bottom of the hill. Eye level is always straight ahead.

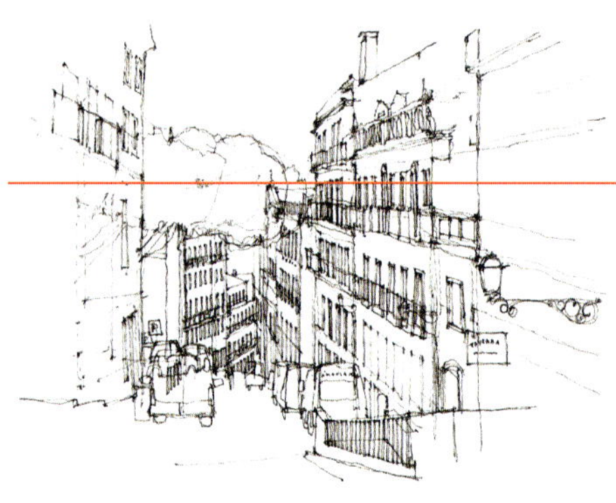

Luís de Camões Square, Lisbon, Portugal

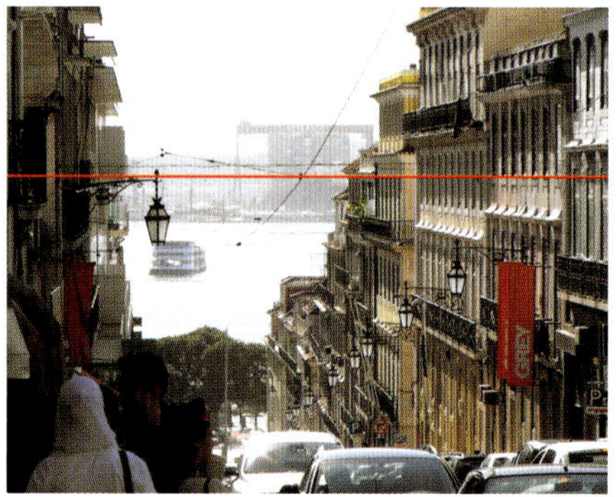

Rua do Alecrim, Chiardo, Lisbon, Portugal

VANISHING POINT

As described earlier in this chapter, the point where objects appear to converge is known as the vanishing point, and it always sits on your eye level.

In one-point perspective, there is a single vanishing point; in two-point perspective, there are two. More complex systems, such as multi-point perspective, also exist, and we'll explore those later. But in both one- and two-point perspective, the vanishing points are always located on the eye level.

It's also important to understand that the vanishing point doesn't just exist in open space. Objects or people in the foreground can obscure it. Therefore the vanishing point is effectively at an infinite distance. In the instance of this one-point perspective sketch of a street in Singapore, the vanishing point is positioned somewhere beyond the van in the foreground.

Take an interior scene as an example – let's imagine we are looking down a corridor with a doorway at the end, as in this photograph of a corridor at the Municipal Hotel, Liverpool. Within the corridor, we can identify the eye level and vanishing point. If the door opens, the vanishing point recedes further but remains aligned with the next doorway – and so on.

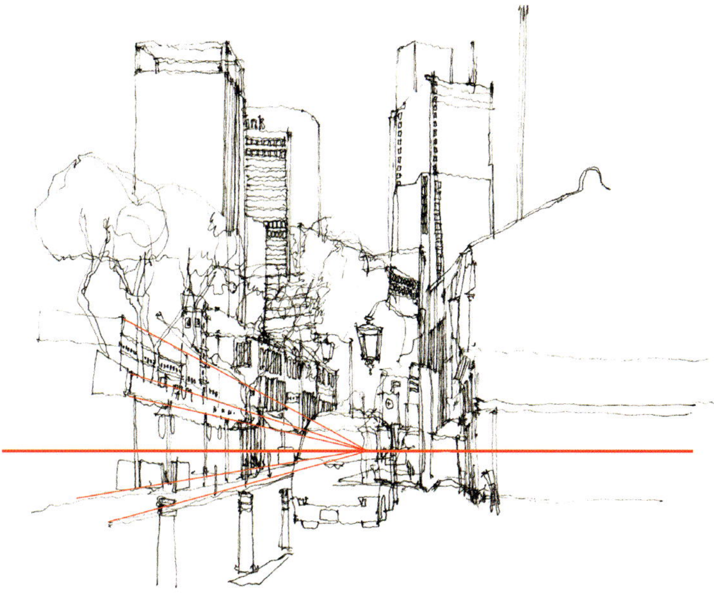

Outside the Thian Hock Keng Temple, Telok Ayer Street, central Singapore

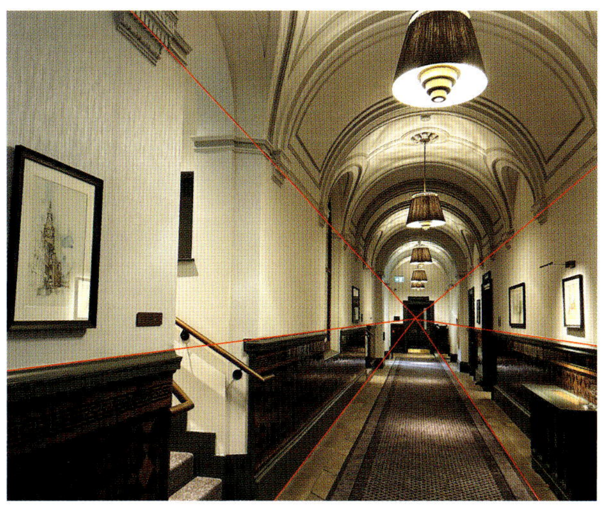

Municipal Hotel, Liverpool, UK

CHANGING THE VANISHING POINT

As discussed earlier in this chapter, if we take a simple view of a street with some buildings on it and we consider a standing-up or sitting-down eye level, the angle of perspective will change to reflect the different eye levels.

The following diagrams clearly demonstrate what happens when the vanishing point changes position. Each diagram represents the same space with a consistent eye level. In diagram (A), the vanishing point is centrally positioned, creating a symmetrical view. In diagram (B), the vanishing point shifts to the left, altering the perspective – receding lines on the left appear steeper, while those on the right become shallower. This effect is even more pronounced in diagram (C), emphasizing how the vanishing point's position influences our perception of space. These are simple principles that should become instinctive (or intuitive) the more you draw in perspective.

Train Your Eye

It can be helpful to take photographs to train your eye to see how lines recede towards the vanishing point. If you hold your camera or phone at eye level and look straight ahead, the resulting photograph will also be at eye level, meaning the principles of one-point perspective described here will apply.

These diagrams illustrate a changing vanishing point – it is important to note that eye level remains the same in each one.

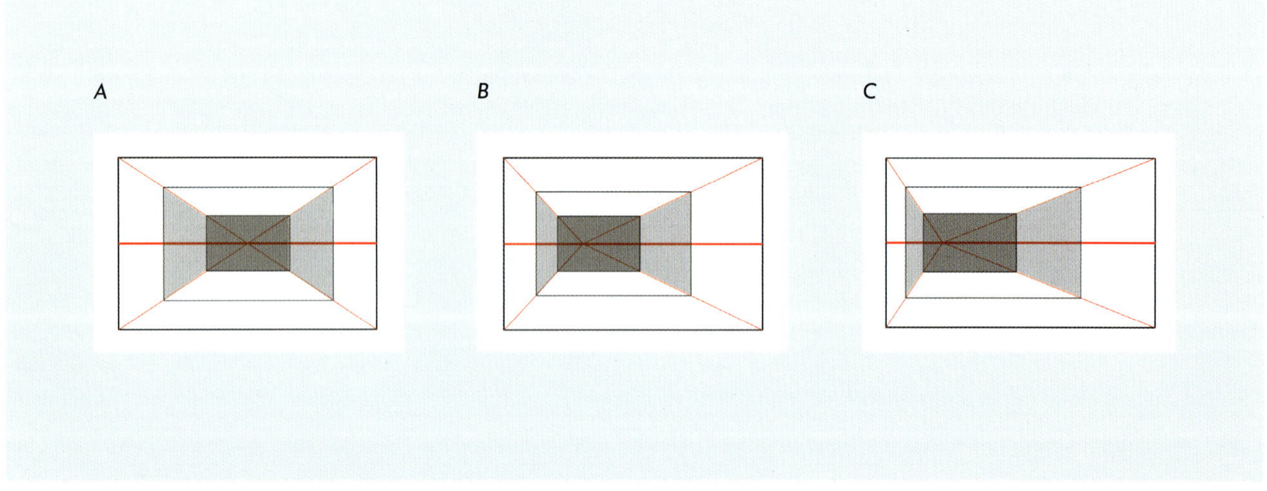

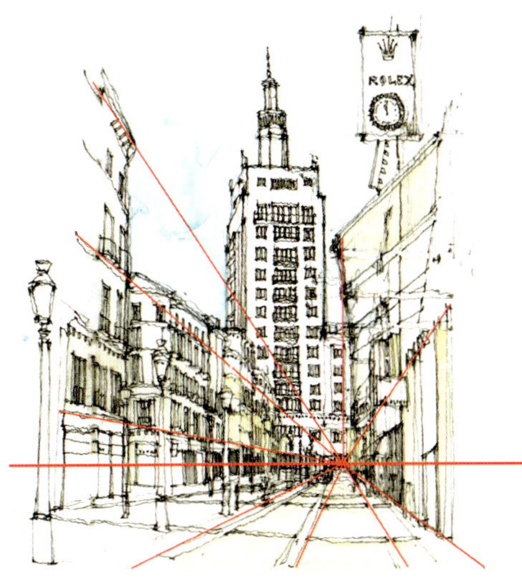

Applying these principles to a real sketch, here is a one-point perspective view of a street scene in Malaga, Spain. The eye level is quite low because I was sitting on a stool while sketching. Since I was closer to the buildings on the right, their converging lines appear steeper towards the vanishing point. In contrast, the buildings on the left recede more gradually, creating a gentler perspective.

FORESHORTENING

Foreshortening is a key part of architectural perspective and as you understand how to draw in perspective you should pick this up intuitively. It's the effect that makes architectural forms appear realistically shortened as they recede into the distance. This photograph of Edinburgh illustrates the point: although the buildings are of similar height, they appear increasingly compressed as they extend into the distance.

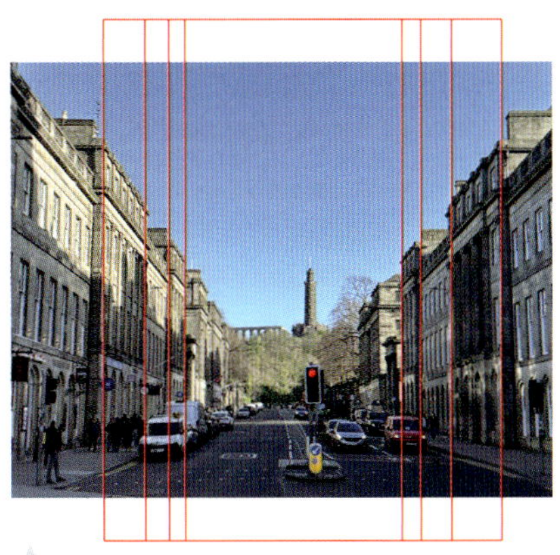

Looking towards Calton Hill, Edinburgh, UK

EXERCISE

Look at these photographs of St Paul's Cathedral, London, and Amsterdam Central Station and see if you can work out where the respective eye levels and vanishing points are.

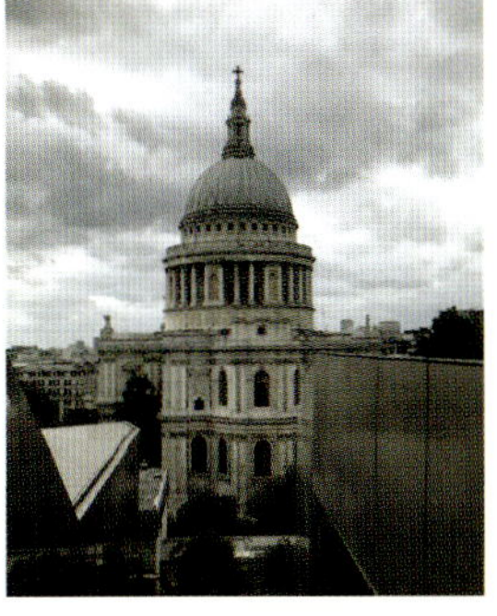

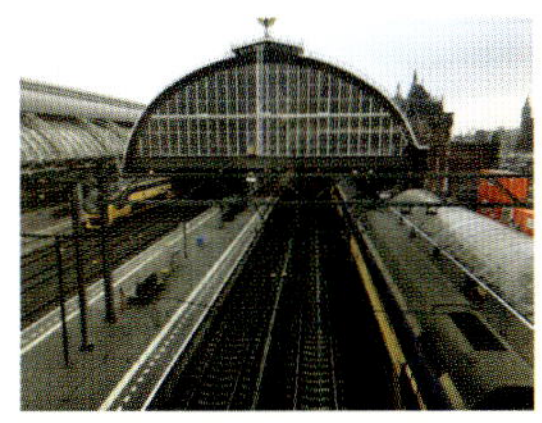

SCALE AND PROPORTION

Understanding the basics of scale and proportion will help you with perspective drawing. And to do this, your pencil is your friend! It serves as a useful measuring tool. For example, hold your pencil with your arm outstretched; by comparing the height of the building to your pencil, you can observe the proportions and start to map this out on your paper.

I usually start by identifying the key verticals on the left and right of the view and how they frame the space. The relationship between their height and the width between them – whether it's a narrow street or a wide road – is crucial for getting the scale right.

Using a pen, held vertically, to measure the height of a building in the distance. The overlay shows that it approximately equates to the height of the pen lid.

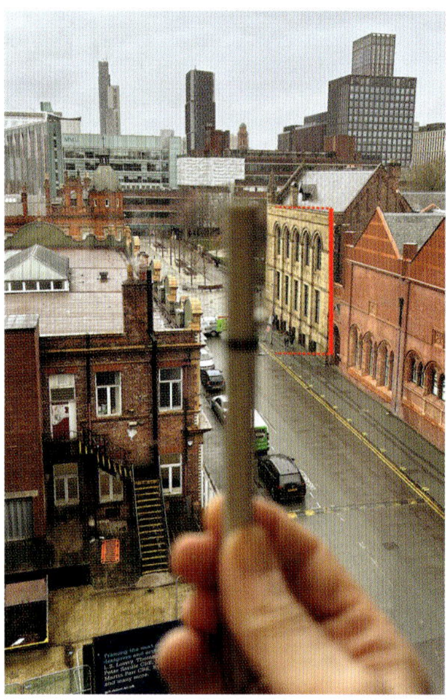

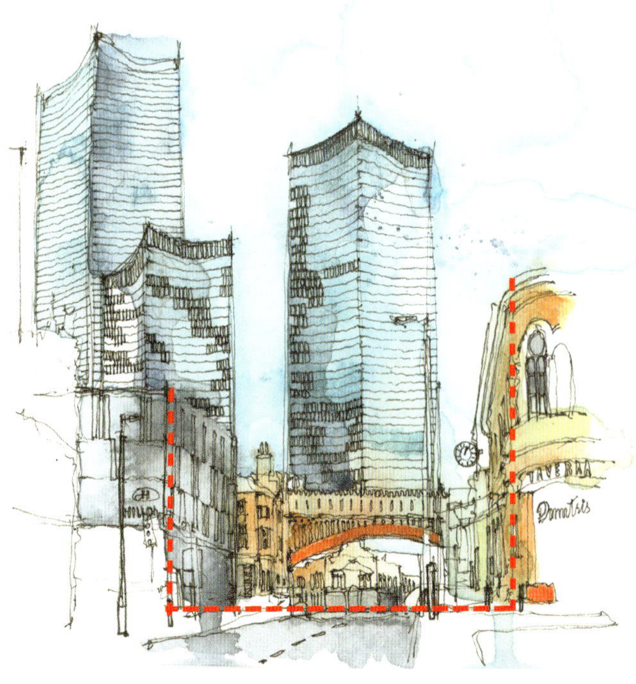

In this sketch of Deansgate Square in Manchester, UK, for example, I had to judge the proportion between the building on the left, the building on the right, and the width of the road, and pavement, between them. You could think of it as a U-shaped space. Here, the road is about one-third wider than the height of the left-hand building, and the right-hand building is roughly equal in width to the road itself. Once those proportions were established, I felt confident the scale of the buildings and therefore the overall composition would sit correctly.

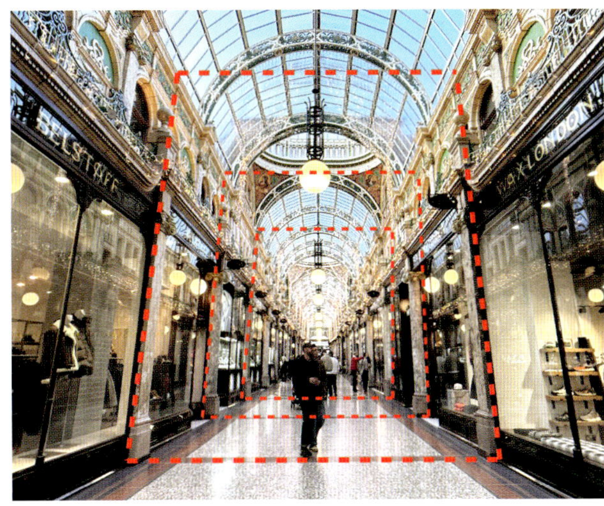

Look at this photograph of County Arcade, Leeds, UK; it's clear the shop units are receding to a vanishing point, but with an image like this it's important to get the proportions right. Think about the width-to-height ratio.

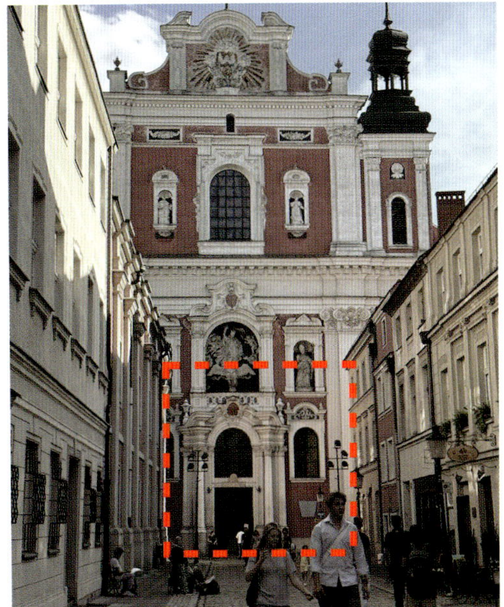

Another example where the proportions really matter: this view is of the Basilica of Our Lady of Perpetual Help in Poznań Old Town, Poland. At the end of this narrow street near the Basilica, the proportion between the width of the street and the height of the buildings is almost equal. As a result, the space formed between the buildings reads diagrammatically as a square.

Tip

I often add figures to my sketches to give a sense of scale and life to the scene. While I don't claim to be particularly skilled at drawing people, they introduce an extra layer of urban interest. In this sketch from Poznań, Poland, the watercolour acts very much as a background layer, with the figures suggested almost as silhouettes in the foreground. I enjoy playing with this balance between positive (painted) and negative (unpainted) space. I also tend to include bags, suitcases, hats, and glasses – which, I hope, help disguise my shortcomings when it comes to drawing people themselves!

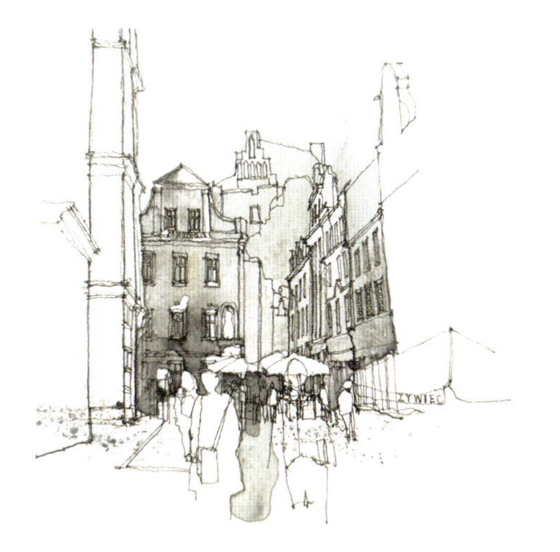

FEATURED ARTIST

Caroline Smith

London, UK

Travel and sketching have always been a huge part of my life, combined with a great interest in urban landscapes and architecture. My preferred drawing technique is a combination of loose ink sketching followed by a watercolour wash.

Cityscapes for me tell a story of how people live and what surrounds them in their metropolitan environment. Skylines are ever changing, places growing and expanding.

My favourite composition is to illustrate a view in portrait format as I love the verticality. This way you tend to end up with a very interesting foreground even though it may not be the focal point. It also allows a layered approach, which I find suits my loose style. "Shade and tone" is my go-to method as it really adds depth and perspective to the artwork without having to go into too much detail.

I usually focus on a key element or landmark and build my cityscape around it. I use soft pencil lines to sketch out the composition, to ensure the scale and proportion between the background and foreground is accurate. Vertical ink lines provide a confidence to the artwork and give that architectural feel to the piece, as well as accentuating the portrait format.

My wash concentrates on picking out three or four key colour tones in the composition to provide a tangible feel to the artwork and allows a connection to the subject and city I'm working on.

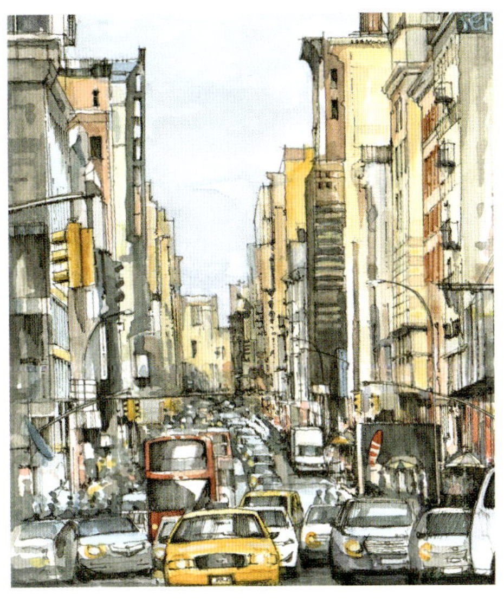

New York City street view

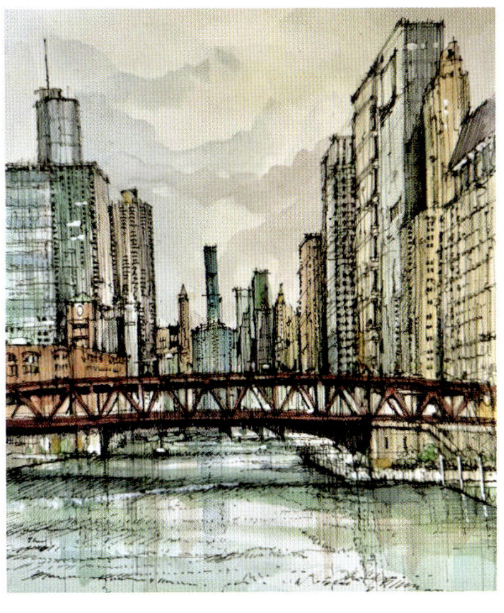

Chicago River, sketched from a boat at river level

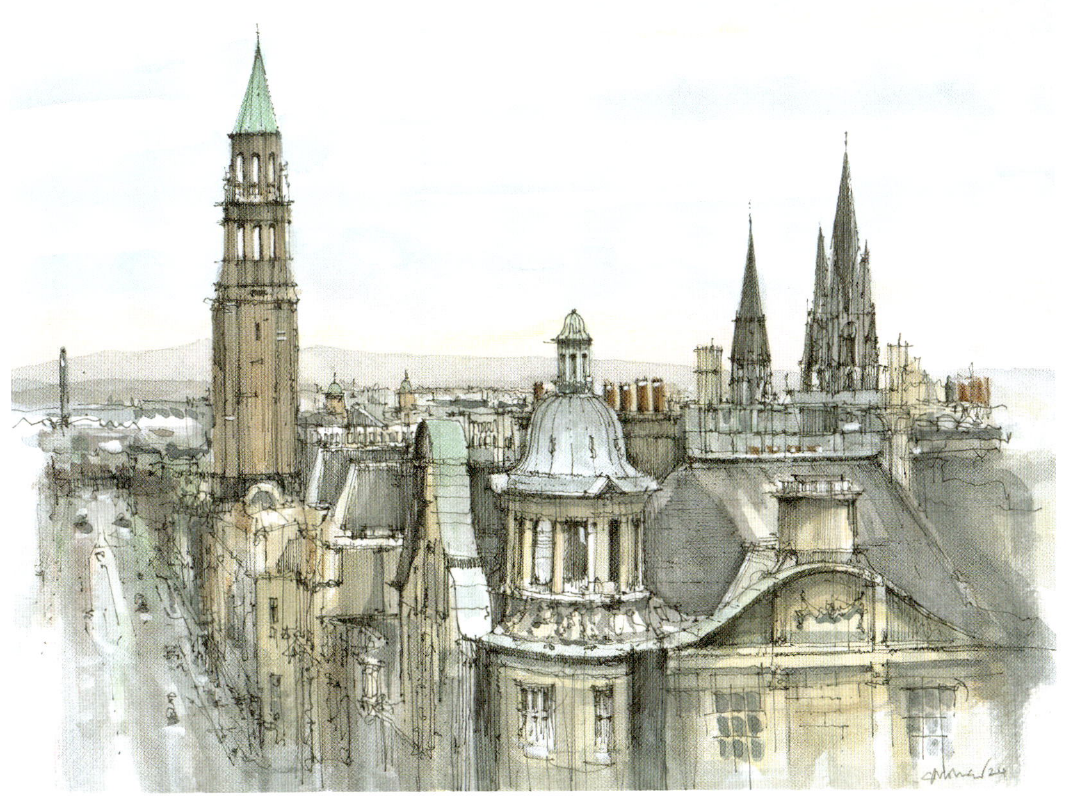

View down Princes Street, Edinburgh, UK, from the Johnny Walker rooftop bar

RULE OF THIRDS

As you become more confident with drawing in perspective and begin focusing on composition, consider applying the rule of thirds – a principle widely used in both art and photography to create more balanced and engaging images.

Imagine your drawing surface divided into a grid of nine equal sections by two vertical and two horizontal lines. Then position your main focus along one of these lines – or ideally at one of the four intersection points – rather than in the centre of the page.

Placing the focus in the left or right third, or towards the top or bottom third, generally creates a more dynamic and visually engaging composition. This offset draws the viewer's eye naturally through the scene and often results in a more balanced overall image.

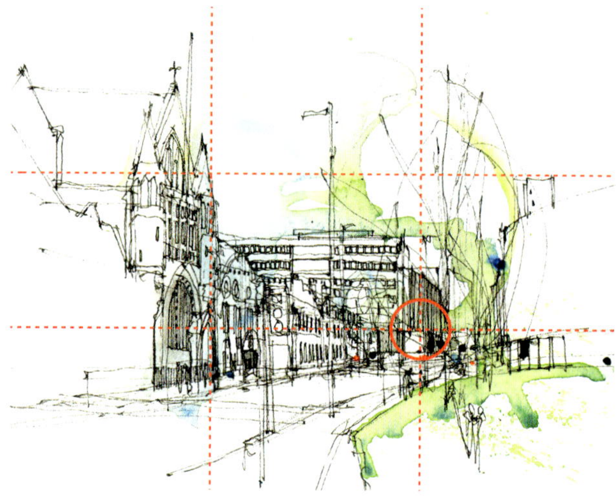

This image shows an overlay of the rule of thirds grid on a sketch of Oxford Road, University of Manchester, UK. You can see the main focus positioned at the intersection of the lower and right-hand thirds.

Note

The rule of thirds does not require a specific drawing size, as it is simply a method for organizing composition. If this concept interests you, you may also want to explore the golden ratio, a more precise compositional system based on the Fibonacci sequence. The golden ratio is applied through the golden rectangle, which is also widely used in architecture to subdivide and organize space.

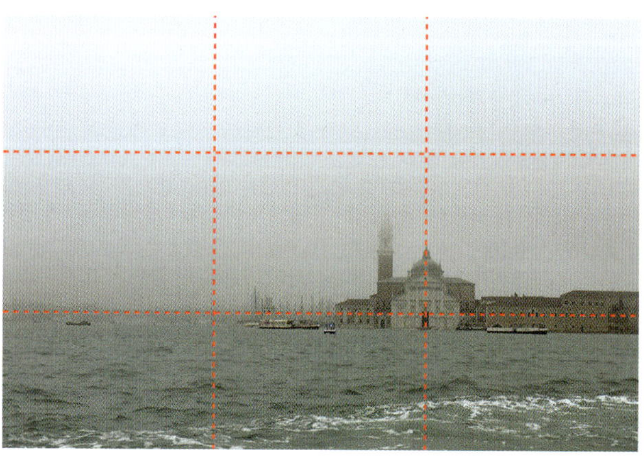

Looking across to San Giorgio Maggiore from St Mark's Square (or San Marco), Venice, Italy

Viewfinders

A viewfinder helps you focus on composition, proportion, and perspective. Using a viewfinder is helpful for beginners and is simple to use (and to make!) or you can buy one from an art materials shop.

Copy your viewfinder's grid approximately onto your paper; then establish the view you want to sketch. You can then replicate the simple shapes of the view onto your paper, aligning with the viewfinder's grid. This is a good exercise that helps with scale and composition. You can adjust the viewfinder to crop in or out, or switch between landscape and portrait formats. Consider the rule of thirds when selecting your composition.

To make a simple viewfinder, cut a rectangle from clear plastic (like acetate) and mark grids of thirds, quarters, or other sizes onto it – both vertical and horizontal – with a permanent marker.

Alternatively, you can use a viewfinder like the one shown here, which allows you to draw directly onto it. This helps you capture composition and perspective on location and then accurately transfer the drawing onto your paper. I bought this viewfinder online.

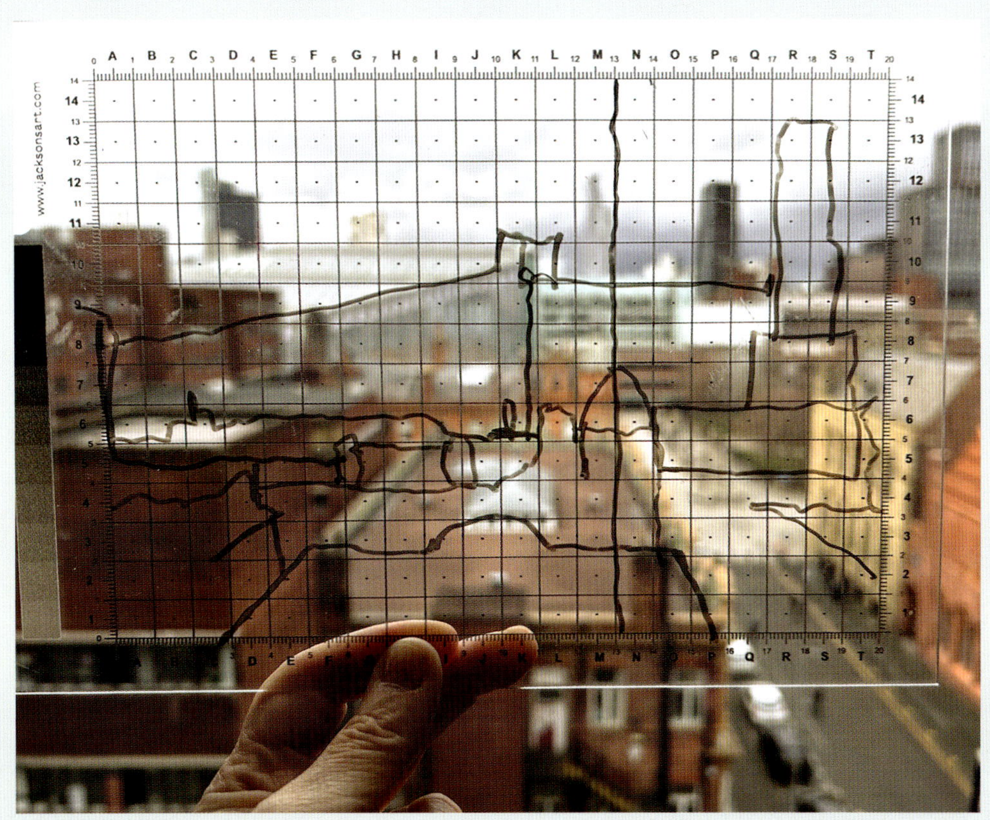

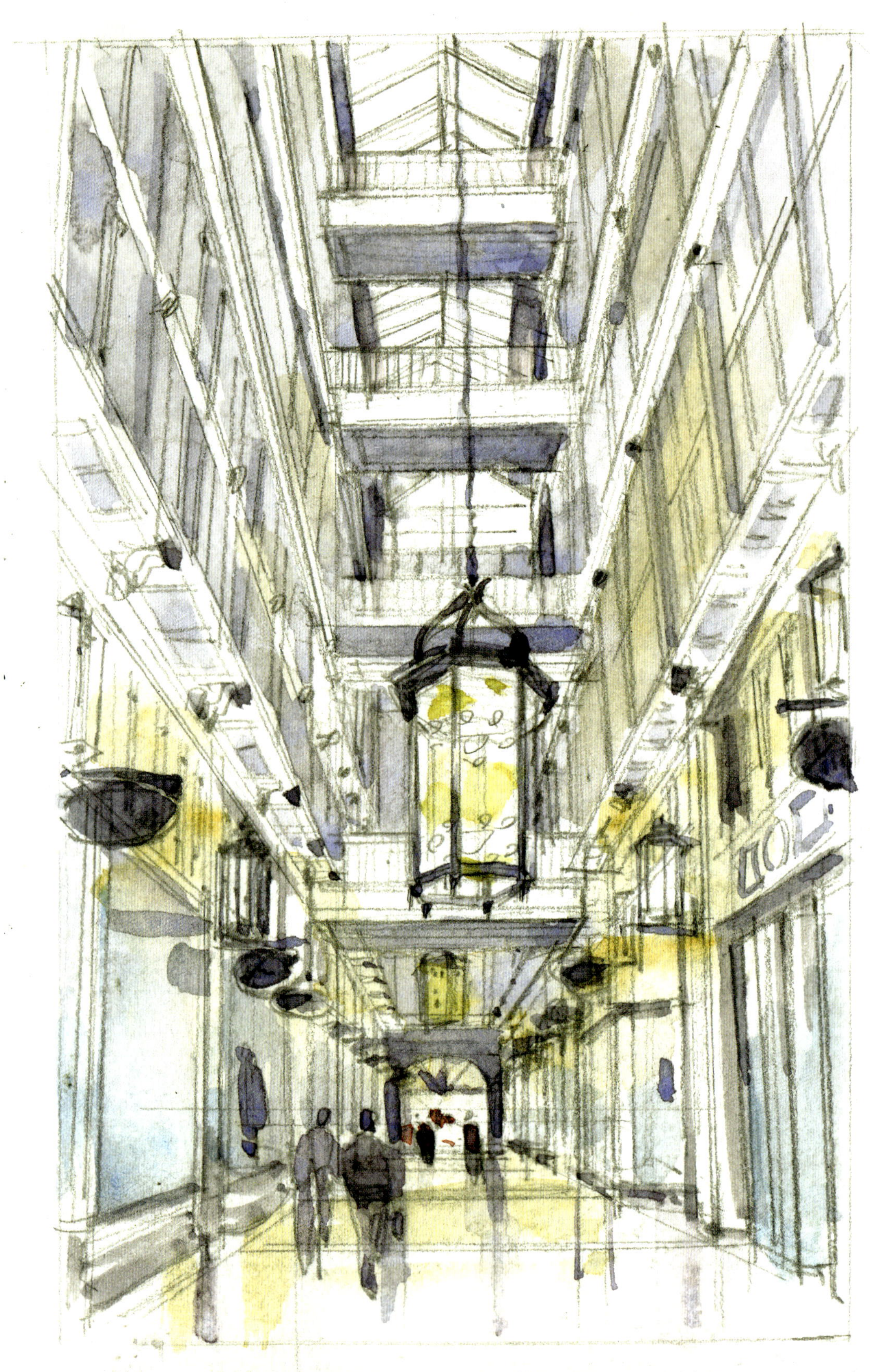

02

ONE-POINT PERSPECTIVE

Now that you understand about the basic principles of perspective – eye level and the vanishing point – let's look more closely at one-point perspective.

One-point perspective is a form of linear perspective that uses specific principles to create the illusion of depth in a drawing or image. It is the simplest type of perspective to learn because it uses just a single vanishing point. It's therefore quicker and easier to set up compared to two-point perspective and other more complex forms of perspective. Mistakes are also easier to spot than with other forms of perspective.

It's an ideal starting point, as it introduces the idea of foreshortening without requiring a complex understanding of perspective. It offers a direct view, making it perfect for views seen straight-on (for example corridors, roads, building façades). It can also create better focus and more drama, as all lines converge to one point, drawing the eye towards the main point of interest.

One-point perspective drawing Strand Arcade, Auckland, New Zealand, by US architect and illustrator Stephanie Bower

INTRODUCTION

First, let's look at the general principles of one-point perspective.

This diagram shows several rectangular objects placed above and below a horizon line. I refer to it as a horizon line rather than an eye level because, in this hypothetical diagram, no actual viewer is implied; the vanishing point is simply positioned to demonstrate the principles. Objects below the horizon line recede upward towards it, while those above recede downward. As a result, we see the undersides of the objects above – as if looking up at them – and the tops of the objects below, as if looking down. Fundamental to understanding this is the relationship each of these objects have with the horizon line.

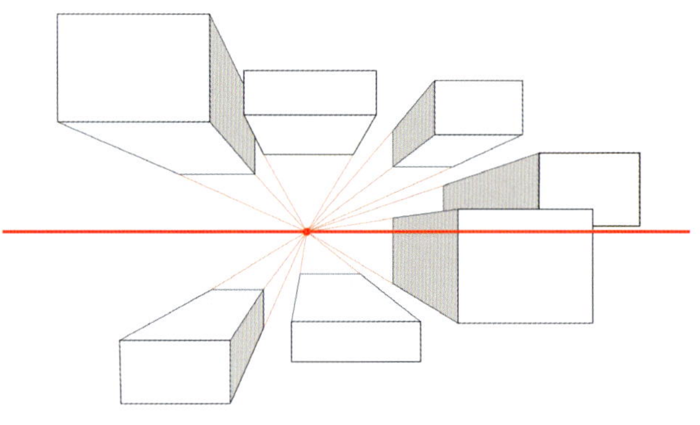

Diagram showing objects above and below eye level in one-point perspective.

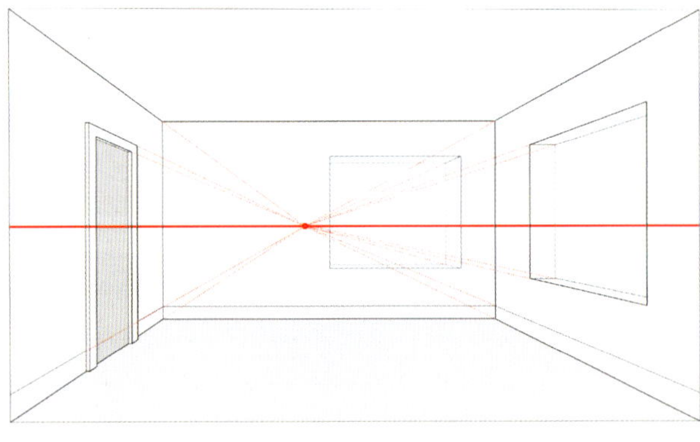

Taking these principles into an interior setting, this image shows a simple room drawn in one-point perspective. The eye level is positioned around the midpoint of the room's height. The wall facing us is shown in elevation, while the side walls – including the window and door – recede towards the vanishing point.

In both the diagrams shown here, there are two planes: one that we see straight-on and the other that is perpendicular to us. The straight-on or "flat" plane is parallel to us, and we would describe this as being "in elevation". The perpendicular one is receding. It's important to note that the rules of one-point perspective apply only to elements in your view that recede towards the vanishing point – much like the telegraph poles described in Chapter 01.

Note that in a one-point perspective view, the viewer must be parallel to the picture plane for the perspective to be correct.

In this sketch of Campo Santa Stefano, Venice, Italy, the building sides parallel to the viewer are marked. All other walls recede towards the vanishing point. Training your eye to see scenes in this way makes perspective much easier to understand.

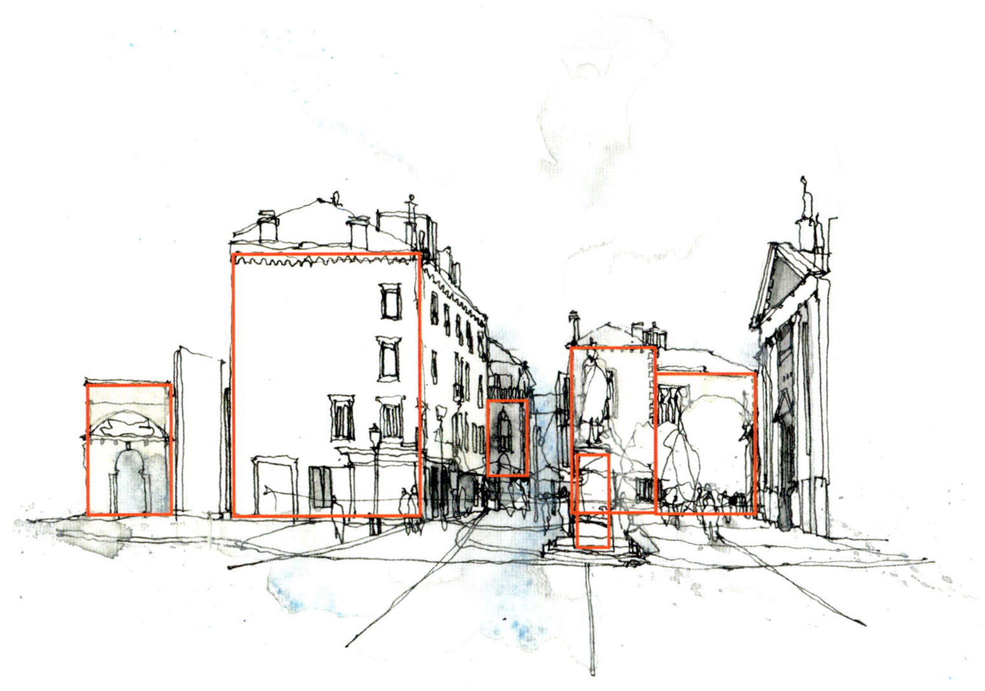

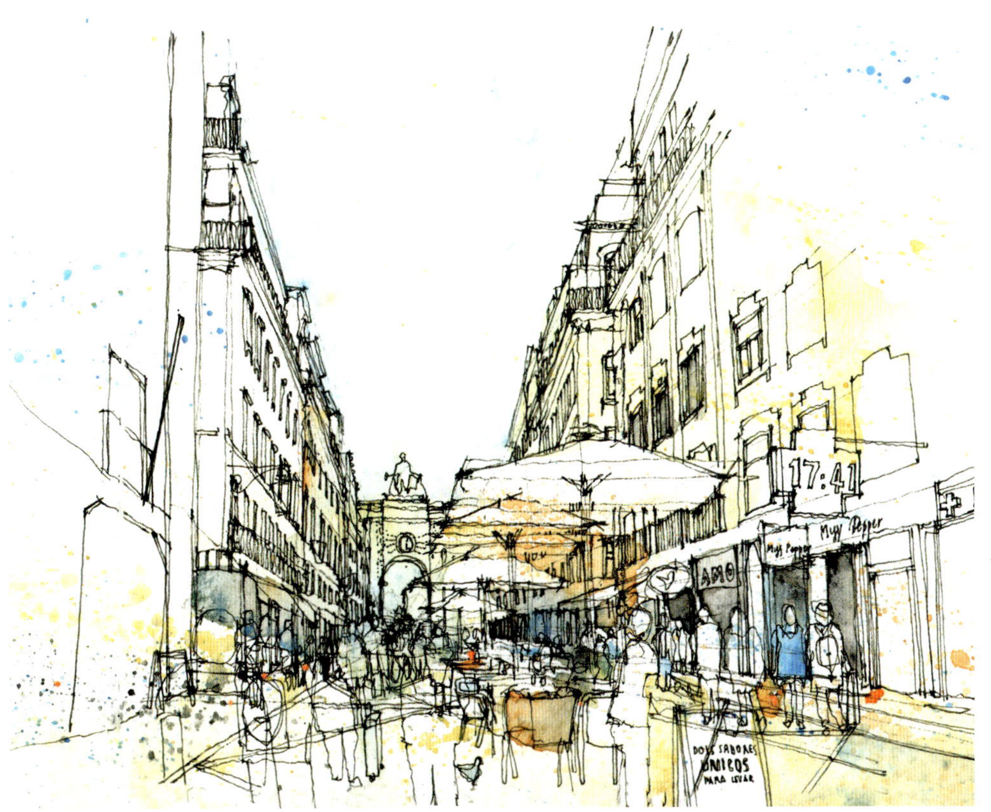

This one-point perspective view of Rua Augusta, the main pedestrianized street in Lisbon, is one of my favourite sketches. I wanted to capture how busy the street felt, with lots of people, signs, and visual clutter, and the watercolour splashes add to that sense of energy and movement.

IN PRACTICE

When I choose what to draw, I usually take a one-point perspective viewpoint. As mentioned previously, the key is understanding what lies above or below the vanishing point.

First, we determine where our eye level sits in relation to the scene in front of us; then we locate the vanishing point. (Sometimes there will be *multiple* vanishing points, something we will look at in more detail in Chapters 03 and 04. For now though, let's assume a straightforward street view where a series of buildings appear to get smaller as they recede into the distance, following the principles of one-point perspective.)

Remember, in one-point perspective, one plane faces us directly, while the perpendicular plane recedes towards the vanishing point. By studying the view, we can identify where receding planes converge, whether from above or below the vanishing point. Through careful observation, you will begin to intuitively grasp the nuances of perspective. This level of understanding comes with lots of practice!

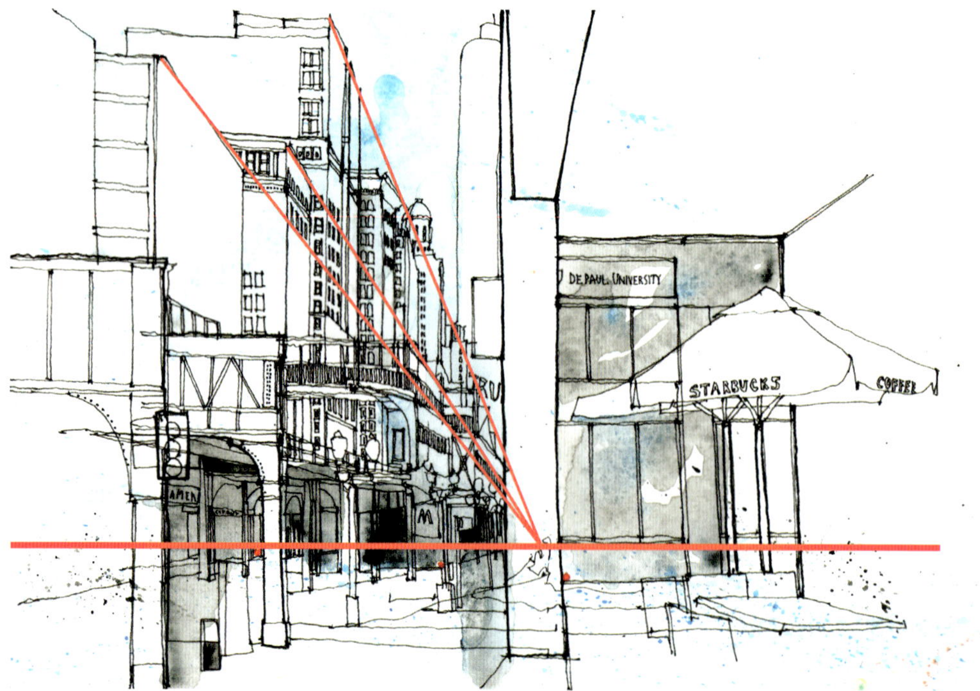

This is a one-point perspective sketch of the Chicago Loop. The sides of the tall Chicago buildings, as well as the elevated Loop railway system, all recede towards the vanishing point, which is hidden behind the column next to the Starbucks umbrella.

Taking this into a one-point perspective street view, let's consider this photograph of a street in Manchester, UK. After establishing our eye level and vanishing point, we now need to determine how much of the view is below eye level and how much is above eye level.

This is actually quite simple to do. This photograph has been taken from a standing position and the vanishing point is central to the view. Therefore not much of the view is below us; if we were to measure it, it would be approximately 1.5m (5ft) at standing, which means our eye level aligns with about mid-ground floor height. However, most of the buildings in this view are at least five storeys in height. As a rough estimate then, about 10% of the building "zone" falls below eye level, with 90% above. Simply being aware of this can help us better structure our composition.

By processing this information, we can understand which elements are parallel to us – appearing as flat planes – and which are perpendicular – diminishing in size towards the vanishing point. This principle applies to everything, including doors, windows, signage, and street furniture. But remember, elements that are parallel to us (flat planes) are not affected by perspective.

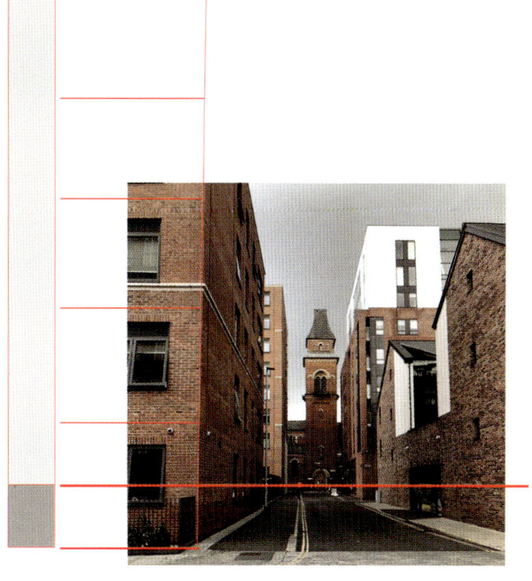

Looking towards Halle St Peters from George Leigh Street, Ancoats, Manchester, UK. The shaded bar shows height proportions: light grey (above eye level) = 90%, dark grey (below eye level) = 10%. As the building has five storeys, each storey is approximately 20% of the total height.

Tip

The same principles apply when sketching a perspective view with very tall buildings – think of a street scene in New York. The relationship between eye level, the ground, and the top of a 50-storey skyscraper might result in only about 1% of the building being below eye level, while the remaining 99% is above it. Understanding this proportion helps maintain accurate perspective and composition.

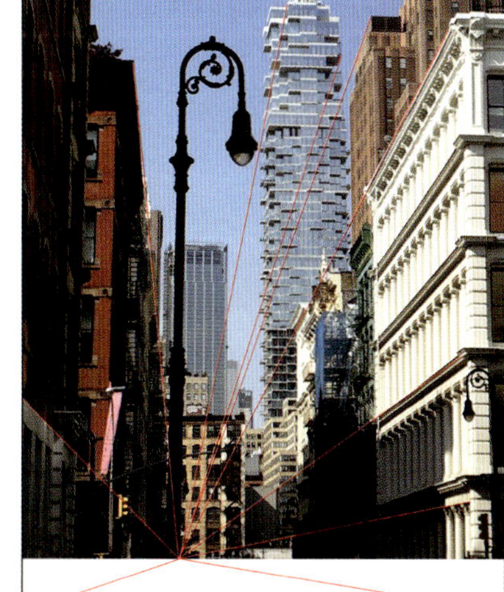

Jenga Tower, 56 Leonard Street, Tribeca, Lower Manhattan, New York, USA

USING AN ELEVATION

Depending on the view you're drawing, it can be helpful to start with a basic elevation. This gives you a solid structure to work from and helps you place the perspective elements of the building more accurately. Also, it's simpler to draw an elevation, especially if you are new to sketching buildings or architecture. Think of the elevation like a template – it helps you map out storey heights, walls, doors, windows, and other key architectural features.

Seeing things in elevation also helps with perspective elements. If a building has windows or doors on a flat plane parallel to us, we can establish a guiding line along the top and bottom edges of a window (or door), extending it to the building's corner. By then drawing this line back to the vanishing point, we establish the window zone in perspective. There are additional perspective techniques to refine this further, which we will explore later.

It's important to note that in one-point perspective we can't always measure depth with complete accuracy; in most cases, we rely on the eyes unless we move into more advanced methods, which we'll cover in Chapter 06. Heights, however, can be established with confidence. For example, a building drawn in elevation in the foreground can provide reliable references for windows, doors, and storey heights.

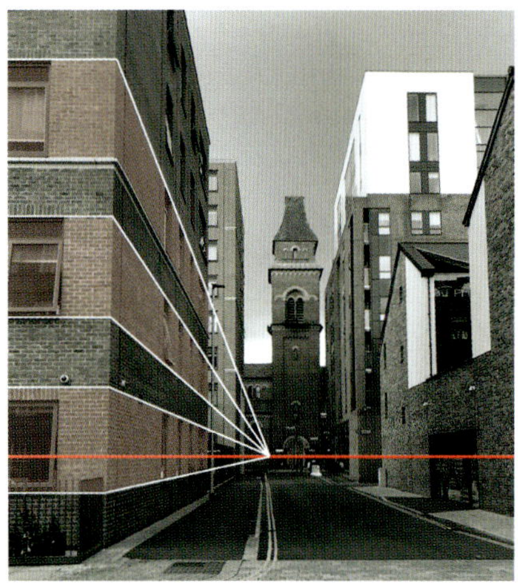

In this overlay (using the image from the previous page), lines indicating the top and bottom of the windows in elevation are extended to the building's corner, and then by drawing guide lines back to the vanishing point, we can establish the top and bottom of the windows in perspective.

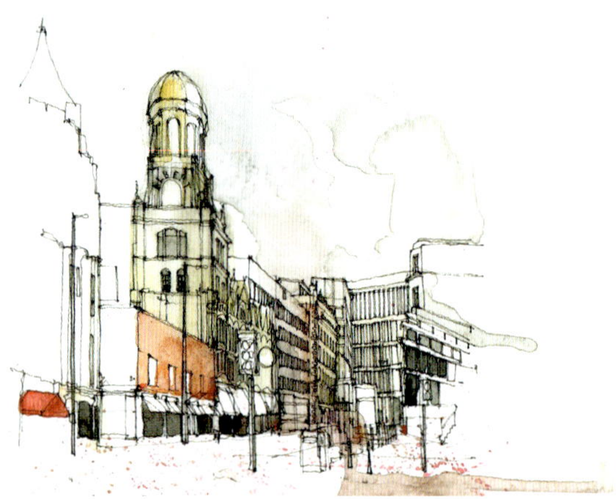

This sketch of Peter Street in Manchester, UK, is a clear example of one-point perspective. Most of the buildings recede towards a vanishing point, except for the Albert Hall tower, which remains parallel to our viewpoint.

In this busy street view drawn in Kuala Lumpur, I focused on a lot of detail, particularly signage and urban clutter. I typically start with faint construction or guidelines, and as I build the sketch, my lines become more confident. The last elements I add are the strong vertical lines, like the lamp post in the foreground. This is another example where the vanishing point is obscured by the busy background.

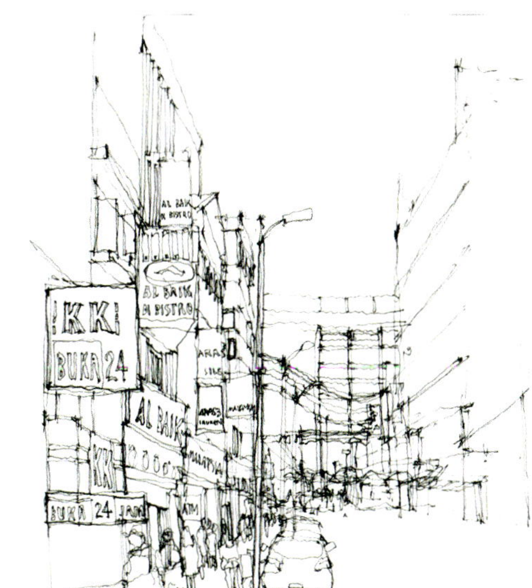

Tip

In addition to using vanishing points to create perspective, overlapping objects is a useful way to enhance a sense of depth.

Using String

If you're unsure about drawing in any type of linear perspective, using a piece of string can be a good way to gain confidence, as this technique ensures that all perpendicular lines recede correctly towards the vanishing point. Setting it up is simple – just fix one end of the string to a drawing pin at the vanishing point. The string can be used like a ruler to help guide perspective lines accurately, ensuring all perpendicular lines recede correctly towards the vanishing point.

This image shows string being used for one-point perspective. For two-point perspective (which we will talk about in Chapter 03), the same concept works: you can use two pieces of string tied together, fixed to drawing pins on each vanishing point.

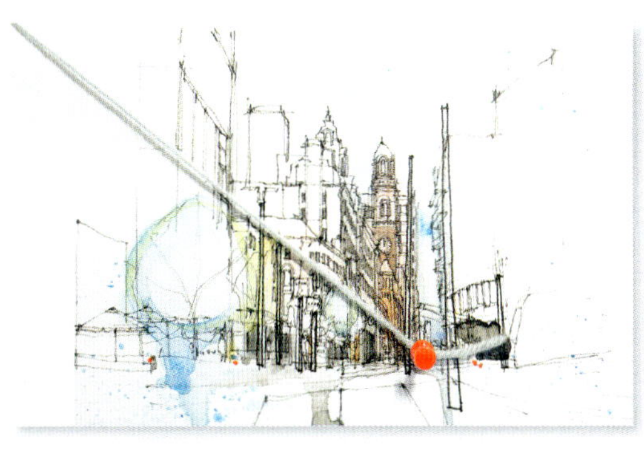

Looking down Oxford Street, Manchester, UK, towards the Kimpton Hotel

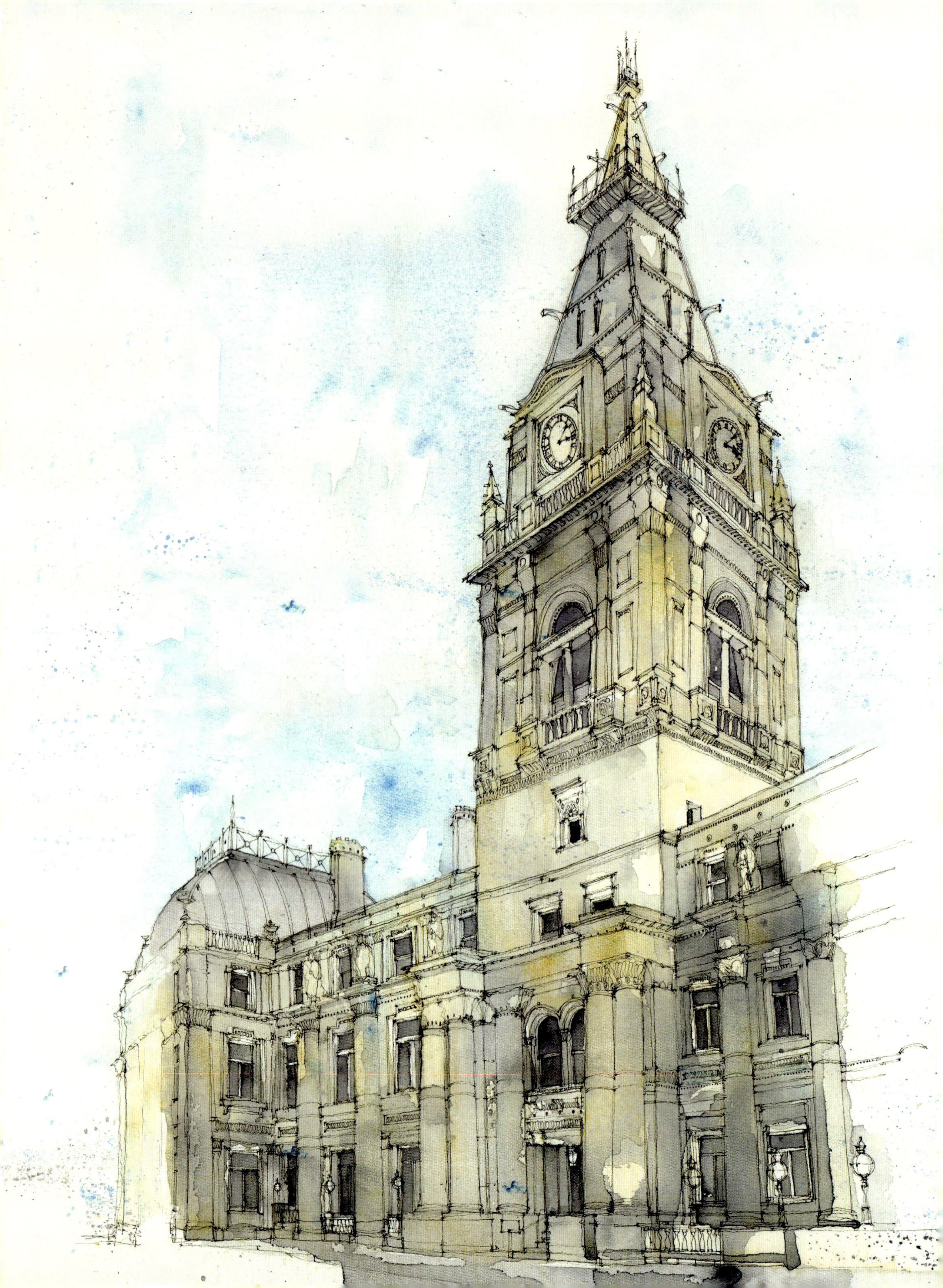

03

TWO-POINT PERSPECTIVE

Two-point perspective is one of the main types of perspective drawing. Unlike one-point perspective, where objects recede towards a single vanishing point (as when looking straight down a street), two-point perspective uses two vanishing points. This allows objects – such as buildings – to recede in two directions, creating a more dynamic sense of depth.

Personally, I think one-point perspective works better for street scenes, but two-point gives you more scope when you're showing angled views. This approach is also more informative, as it allows you to see two sides of a building or object at the same time.

For on-location sketches, a basic understanding of two-point perspective principles is usually sufficient, but for more professional visualizations, a thorough grasp of these principles is more important.

A two-point perspective illustration of the Municipal Hotel in Liverpool, UK. Commissioned to celebrate the opening of the new five-star hotel.

Two-point perspective can be a little more challenging to work out than one-point perspective, but the principles are still the same. It follows the same basic rules – we still need to figure out where eye level is – but instead of just one vanishing point, we have two.

Two-point perspective is an effective way to create a more dynamic and interesting three-dimensional view and is widely used in architectural illustration. It is especially useful for depicting buildings, objects, and interior spaces, as it provides a more "active" viewpoint than one-point perspective.

INTRODUCTION

There are two types of two-point perspective set-ups as you can see in the external and internal corner diagrams shown here – both set out from a strong vertical "leading" edge. This can be an external or internal leading edge. While the rules are the same for both, it is important to understand which type you are drawing.

The first one usually relates to an external view where we are looking at the corner of a building (or object) – this is closest to us and the two sides of the building will each recede to a vanishing point (VP). This is more realistic and closer to how we see things in everyday life, rather than the more composed, "set-up" one-point perspective view, where the viewer must be parallel to the picture plane for the perspective to be correct.

Alternatively, when viewing an interior space, the perspective often extends forwards from an internal corner, with the vanishing points effectively hidden out of view. This approach is especially useful in interior visualizations, as it gives clients a clearer understanding of the spatial design than a one-point perspective typically provides.

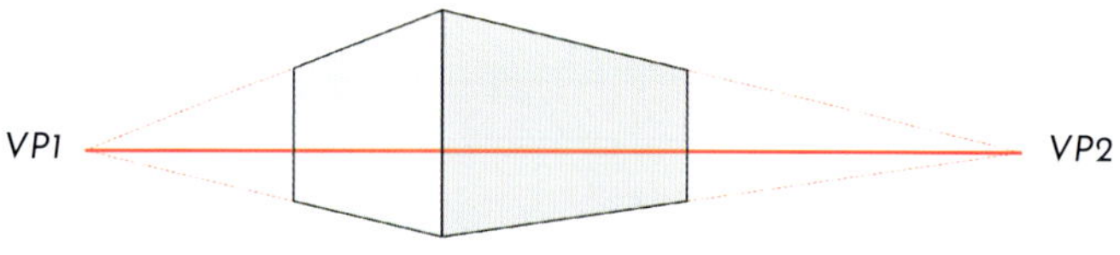

External corner

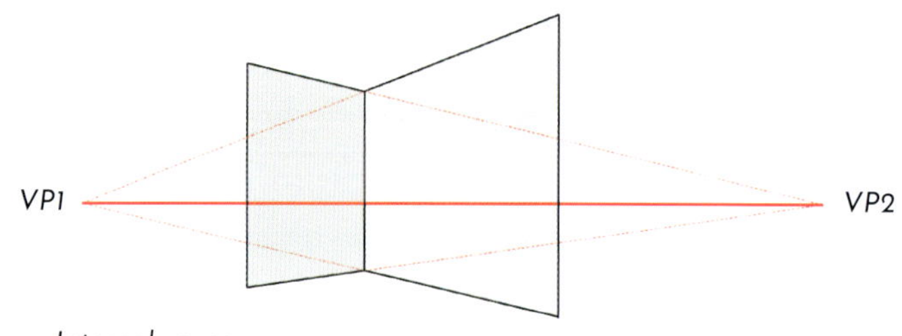

Internal corner

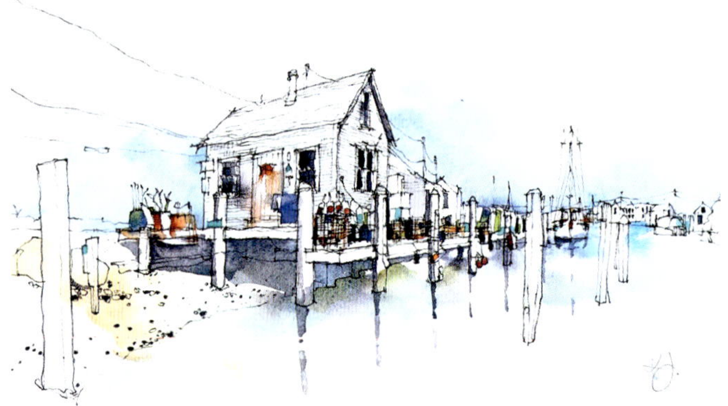

Karen Jones, the artist of this sketch, describes it as "a tranquil meditation sketch (this shack probably featured in the Jaws movie!)", but what stands out to me is how effectively it demonstrates two-point perspective. The two visible walls of the shack recede towards separate vanishing points, while the telegraph wires and posts in the water loosely align with the right-hand vanishing point. Despite its loose, unstructured feel, the perspective is completely accurate.

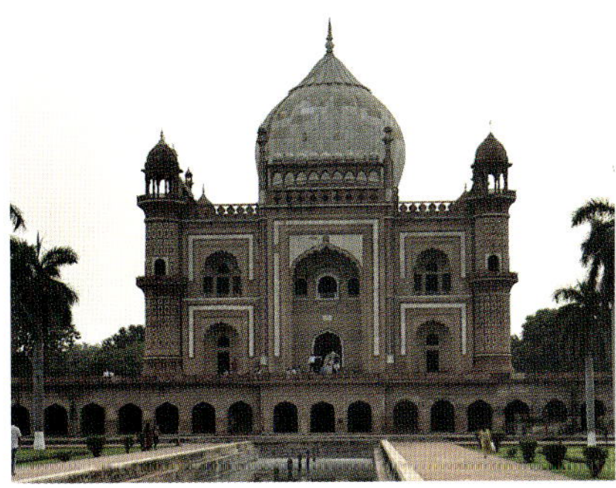

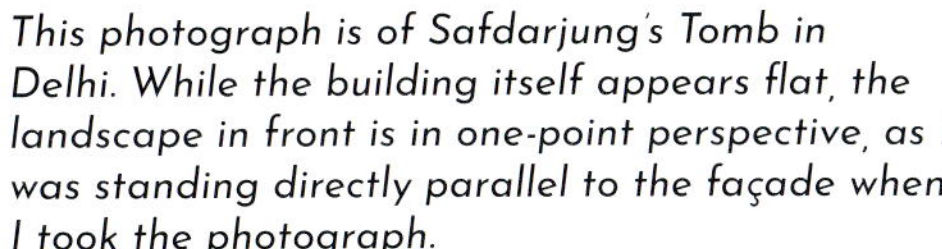

This photograph is of Safdarjung's Tomb in Delhi. While the building itself appears flat, the landscape in front is in one-point perspective, as I was standing directly parallel to the façade when I took the photograph.

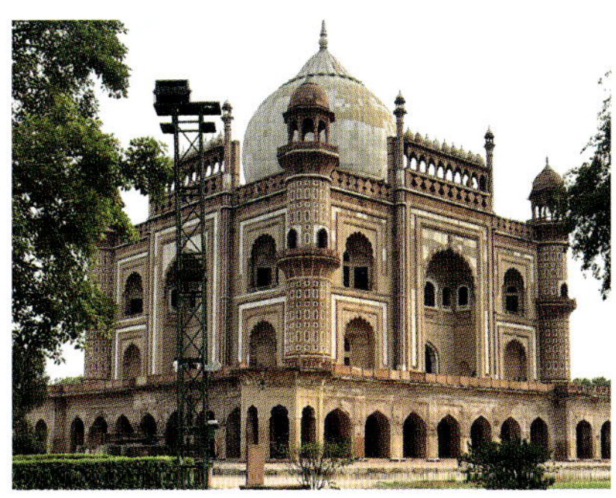

In contrast, this second photograph of the same building captures two sides of the structure. This two-point perspective gives us a much better sense of the building's volume and form, allowing us to appreciate its three-dimensional qualities.

Simple Boxes

When learning to draw in two-point perspective, you can start quite simply by drawing boxes or basic shapes – the architectural detail can follow later.

In (A) you can see one yellow box that has been placed below eye level. We know this because we can see the top of the box, and the two visible sides of the box each recede to a vanishing point located on the eye-level line.

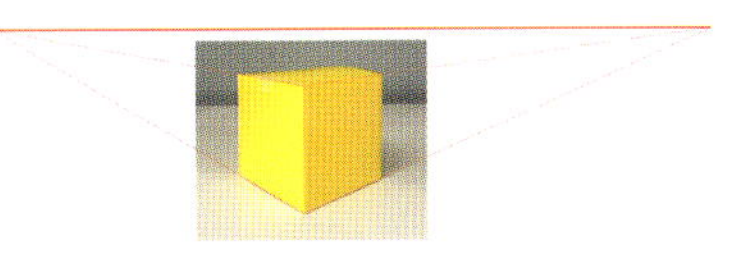

A

In (B), three yellow boxes are stacked, with the bottom box positioned similarly to the single box in (A). This time, eye level sits in the lower third of the middle box. Above this, the box edges slope towards the vanishing points, with the top box showing the steepest downward angles.

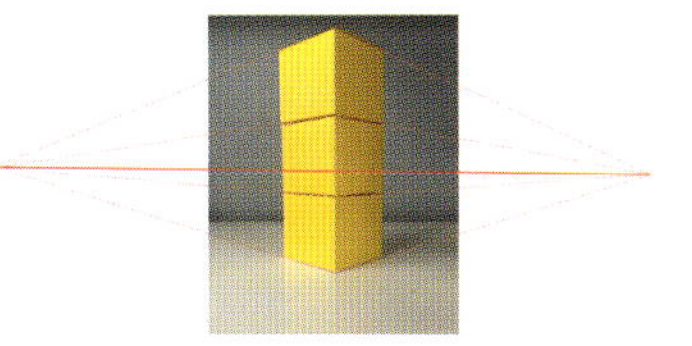

B

In (C), I've arranged a few boxes in different colours. The lower boxes, as in (A) and (B), sit below eye level, while the others are placed above it. It doesn't take much imagination to see these as building blocks, helping you visualize how two-point perspective works in a real-life setting.

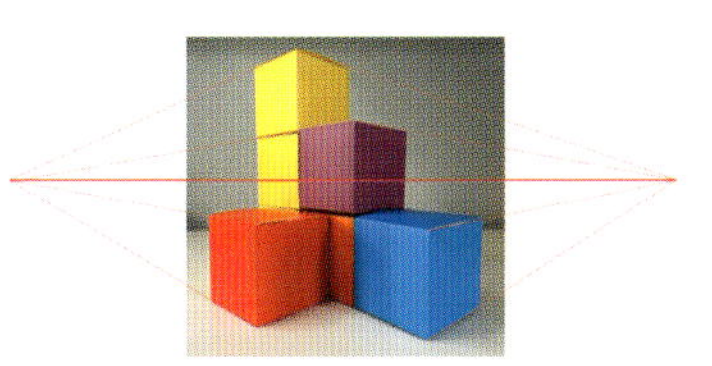

C

IN PRACTICE

As with one-point perspective, we still begin with finding our eye level. We then establish a strong vertical leading line to start setting up the view. Because we are looking for two vanishing points, this vertical line needs to be an external or internal corner.

The same principles apply to both, but as mentioned previously, the outcome is different. When we look at the corner edge of an object or building, it will appear to recede – and slope up or down towards two vanishing points. This helps define the object as a three-dimensional structure. Conversely, using an internal corner in a room to set out our perspective view, we get the reverse situation, and we bring our lines forwards from the internal corner.

It is worth noting that in a two-point perspective view, at least one vanishing point may be positioned outside of the picture plane or paper space. Generally, when a vanishing point is far beyond the edge of the paper, the receding perspective tends to appear shallow.

You can often figure out this perspective naturally as you draw, since the amount of recession is so slight it can feel almost like drawing an elevation (a straight-on view). Alternatively, if you want to set up a more precise two-point perspective view then you could use the string method described in Chapter 02 to work out the angles of recession to this vanishing point.

EXERCISE

Take a look at this photograph of the Book Passage in Sausalito, California. It shows a charming single-storey building in two-point perspective. Your challenge is to imagine how the building might look with an extra storey added.
First, identify the vanishing points to establish your perspective – then get creative with your architectural design!

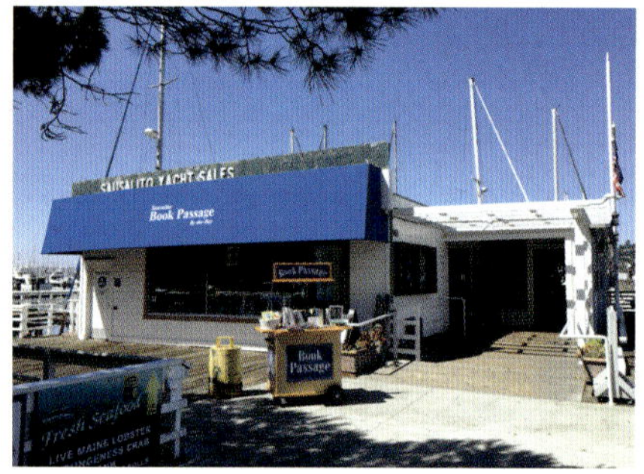

Tip

In two-point perspective, when you're close to a building, the sides of it seem to angle away more sharply because the vanishing points are closer together. But as you move further away, the vanishing points spread out, and the sides of the building look more straight and less angled.

Tall buildings work especially well in two-point perspective, often drawn at a 45-degree angle to highlight both visible sides equally. You can see this in the illustration of the Municipal Hotel in Liverpool shown at the start of this chapter.

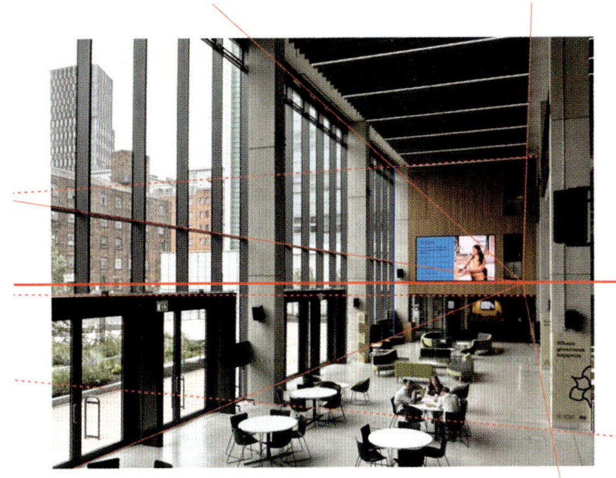

This photograph of the atrium space at Manchester Metropolitan University's Faculty of Science and Engineering, is a great example of a two-point interior perspective. The key internal corner is where the external glazing meets the timber clad wall. Eye level is relatively high as I was standing on a staircase, almost at first-floor level. The right-hand vanishing point is visible, while the left-hand vanishing point lies outside the frame. I have indicated this by using dotted lines that recede gently towards the distant vanishing point.

This photograph of Manchester Art Gallery is a good example of two-point perspective. I was standing on the pavement opposite so my eye level is quite low; you can clearly see the two sides of the building receding to two vanishing points.

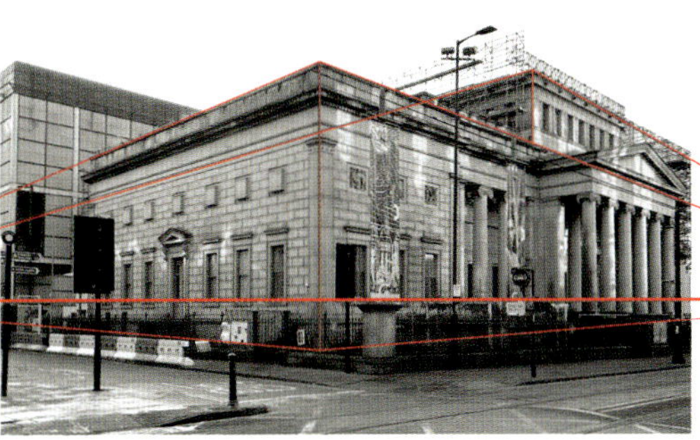

Consider the Clock Face!

Consider this example of Moorlands House in the heart of Leeds' retail district, UK. Focus on the leading external corner and the highlighted projecting cornice, which slopes down towards the vanishing points. This slope can be understood as an angle – try visualizing it as if it were the hands of a clock.

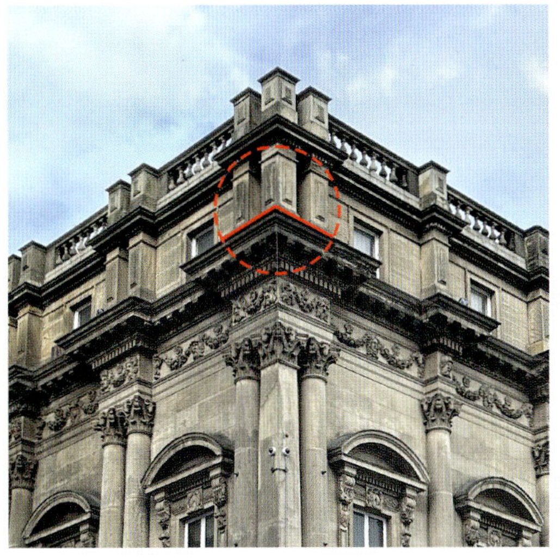

FEATURED ARTIST

Stephen Travers

Sydney, Australia

I think one of the big problems facing anyone wanting to learn perspective is that most free online videos are based on material originally produced for architects and drafting needs.

It's more suited to a simplified, somewhat theoretical presentation than explaining a real-life scenario with uneven ground, curving streets, and objects randomly angled to each other. And the presenters keep secret that what they are saying only works as presented in those quite limited circumstances. They don't tell us what they haven't told us, so we naturally think we've understood the topic when we've understood the video and then blame ourselves when we struggle to apply it all when we draw from life.

If we do instead draw from photographs or life, we could just sharpen our powers of observation for our drawing purposes, and draw what we see, rather than worrying about guidelines and vanishing points at all!

Personally, I like to identify the patterns that perspective creates in the theory diagrams and then look for those patterns in real life. My theory knowledge makes them easier to spot and therefore use them more accurately in my drawing. I feel like it's the best of both worlds and easier than either one of them. Let's all reclaim our confidence with the word "perspective" from the unhappy experiences earlier presentations have given us.

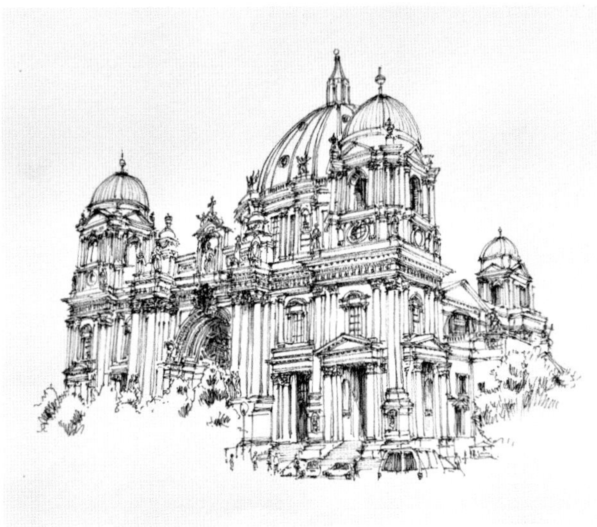

Berlin Cathedral, Berlin, Germany

Vienna State Opera House interior, Vienna, Austria

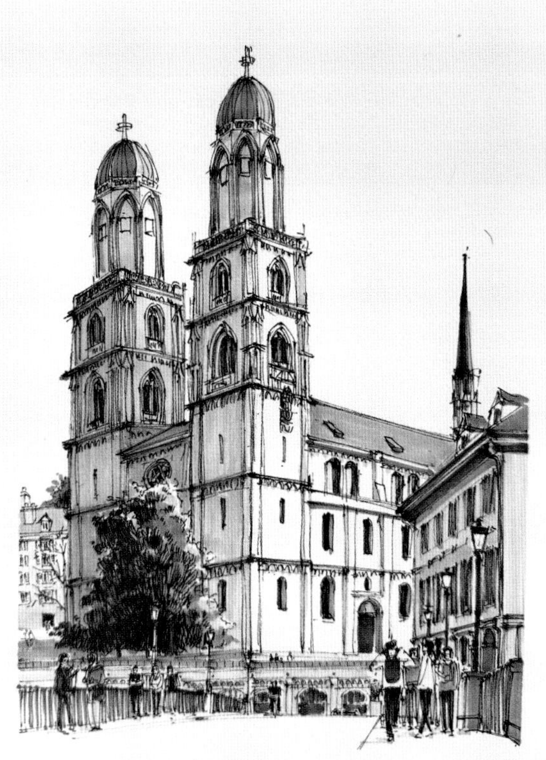

Grossmünster Church, Zurich, Switzerland

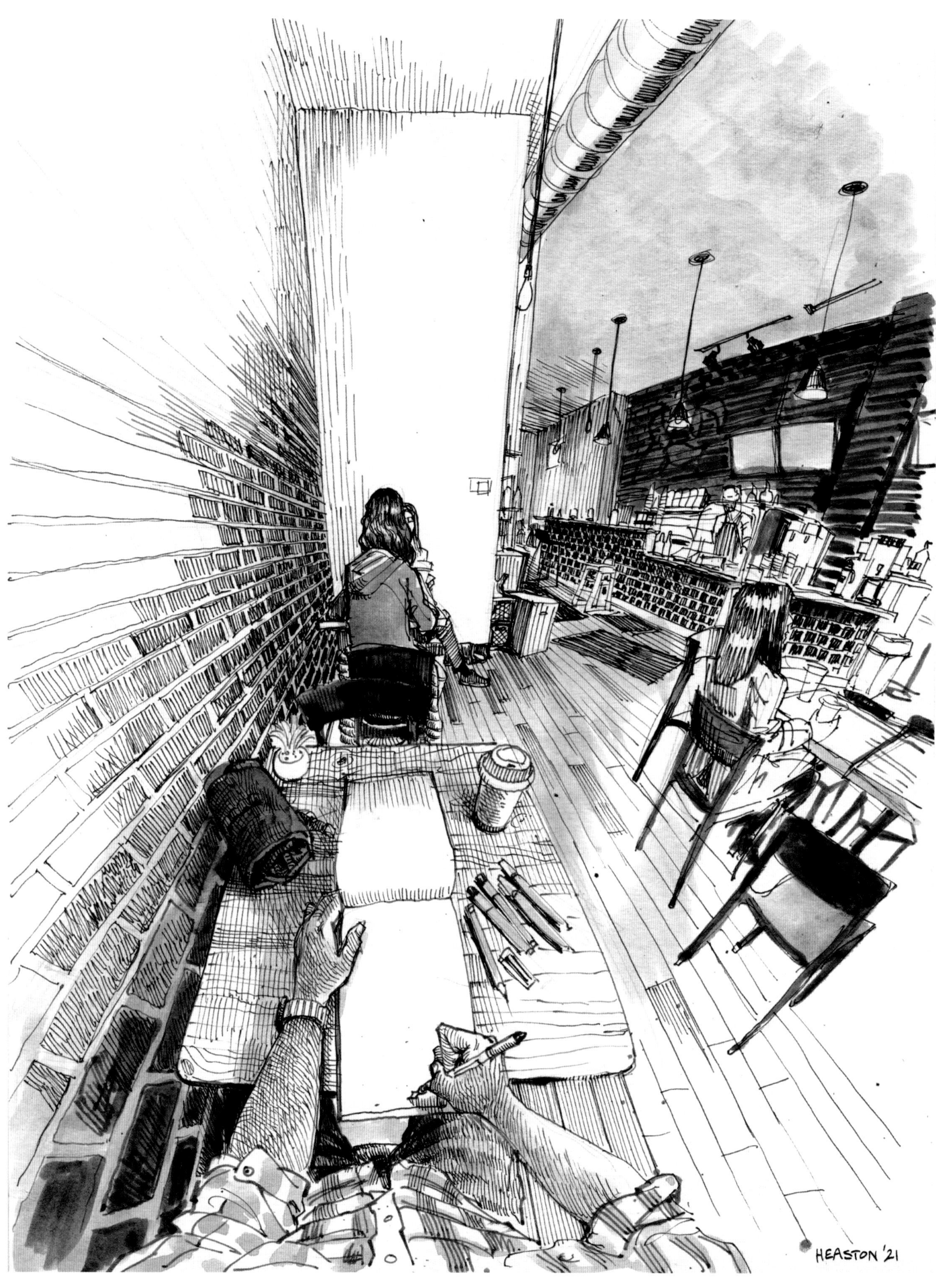
HEASTON '21

04

FURTHER APPROACHES TO PERSPECTIVE

In addition to the familiar one- and two-point perspective techniques, there are several other approaches to consider, which we will explore in this chapter.

As with one- and two-point, these techniques aim to represent three-dimensional space on a two-dimensional surface. In practice, however, what we see in real life is often much messier and does not always fit neatly into a standard one- or two-point viewpoint. This is where multi-point perspective comes in, or using three or more vanishing points to capture more dramatic or unusual views.

It is also important to consider aspects such as shadows and reflections. These, too, follow the rules of perspective and must be constructed carefully in order to appear convincing. In this chapter, we will move beyond the basic principles and begin exploring ways to represent more complex approaches to perspective.

For example, this deceptively simple sketch by Paul Heaston is a masterpiece of complex perspective drawing. The viewer is immediately drawn into the scene, as Paul includes himself within it – you can see his sketchbook in the foreground, and everything radiates out from here. I especially admire his careful attention to detail and texture, as well as his clear, expressive lines. Paul is also very skilled at including figures in his work.

◀

Cherry Bean Coffee, Denver, by Paul Heaston

ATMOSPHERIC PERSPECTIVE

Atmospheric perspective uses a more painterly means of expressing distance, rather than relying on the formal principles of perspective. It is also known as aerial perspective (not to be confused with bird's-eye perspective, which is taken from an aerial viewpoint).

We touched on atmospheric perspective in the Prologue but here we will look at some more contemporary examples. As a reminder, atmospheric perspective is when the depth in a view is established by showing objects that are further away seeming to fade into the distance, with much less detail, whether this is line work, tonal value, hue, or saturation of colour. Objects in the furthest distance often take on the colour of the background, which might be the sky, and so would appear as shades of pale blue, or even reds and oranges if around sunrise or sunset. I should reiterate that this is an artistic technique that is an intuitive response to conveying perspective.

This sketch by UK architect and urban sketcher Karen Jones, beautifully conveys atmospheric perspective in a London skyline sketch. Vibrant watercolour and architectural detail have been carefully applied to foreground buildings, but the watercolour and line quality diffuses into the distance, where the silhouetted skyline fades almost to the colour of the sky.

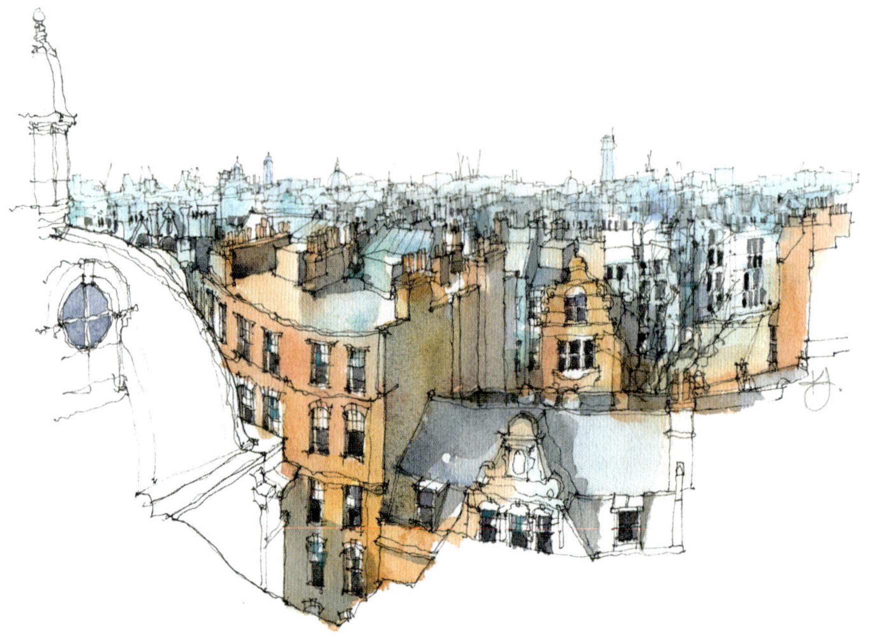

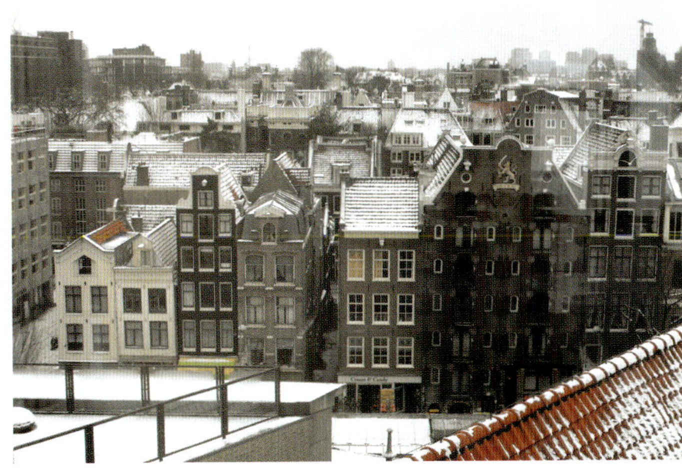

This photograph of Amsterdam convincingly conveys atmospheric perspective. It captures a winter's day, with colours fading into the distance to the point of desaturation as they meet the sky.

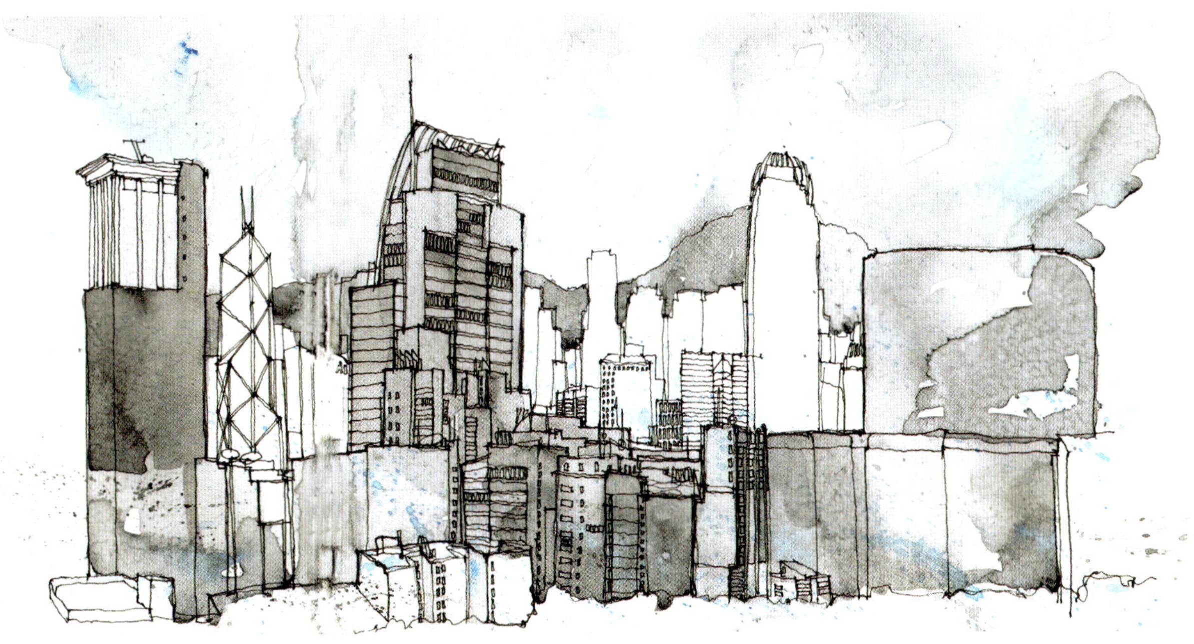

This skyline sketch, which I drew while visiting Hong Kong a few years ago, loosely illustrates atmospheric perspective. Notice how the colours fade, and the details soften as the view recedes towards the hills in the background. This example shows that you don't always need to follow the rules of perspective rigidly, particularly when sketching on location. Instead, you can apply them more loosely – provided you understand the principles.

FEATURED ARTIST

Darman Angir

Surabaya, Indonesia

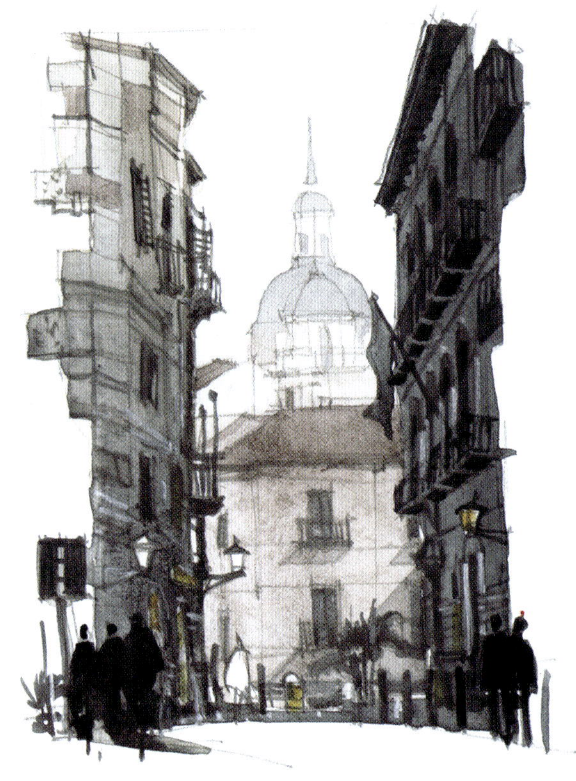

San Isidro, Madrid, Spain

Perspective drawing certainly shows us visual depth, but we can put greater pictorial impact when spatial layering techniques are added. Using more detailed drawing on the foreground layer and less on the background layer is one of those techniques, highlighting the distinction between each spatial layer and aerial perspective of distance.

I always think the whole sketching process consists of two major parts: the drawing and the colouring. Both have their own merit; personally I see the drawing part as more essential and harder to learn. Especially in perspective drawing, even when we understand all the technicalities of it, experience and intuitive adjustment will finally guide us to make a more visually pleasing drawing.

There are numerous compositional options available, yet I find the rule of thirds one of the most helpful, particularly for asymmetrical compositions.

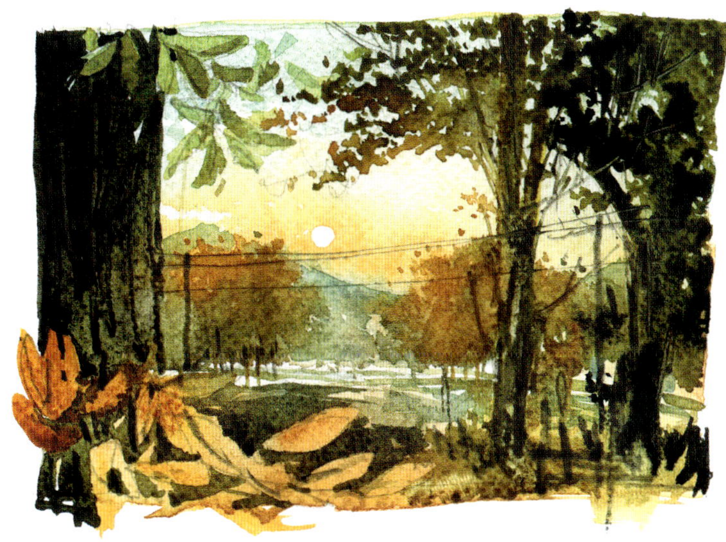

Singhasari, Malang, Eastern Java, Indonesia

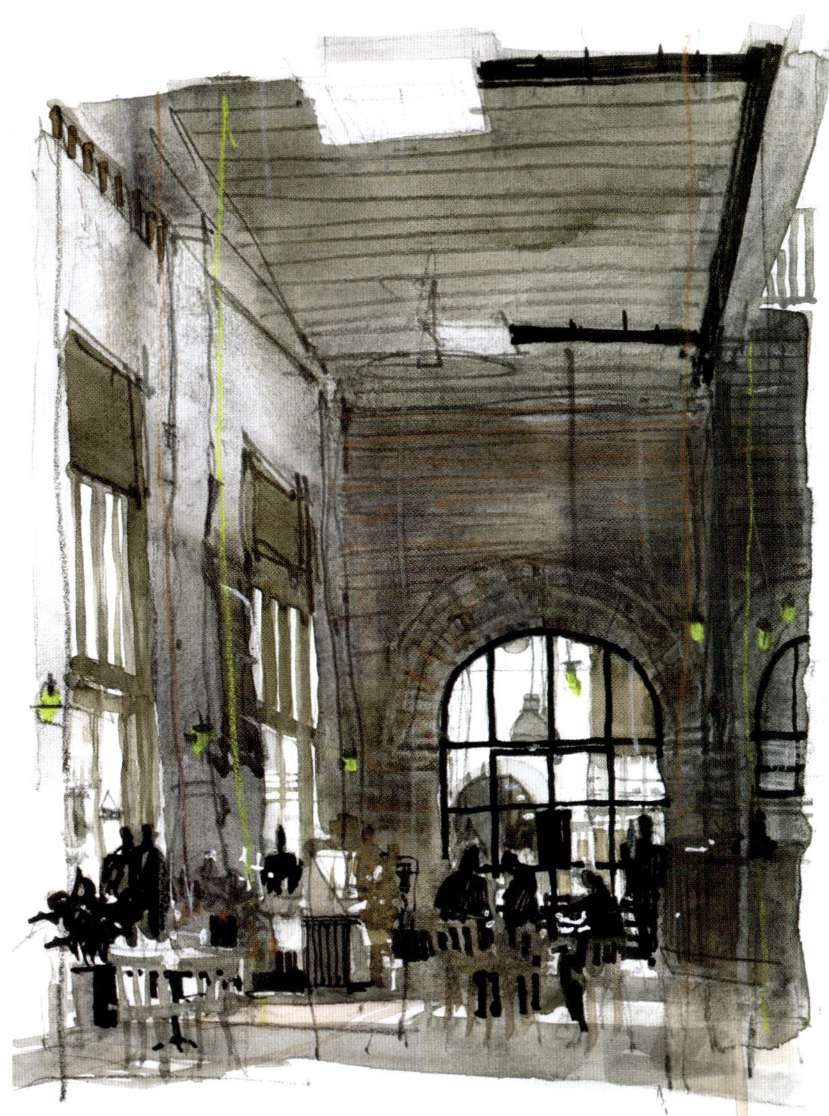

Spiegal Bar and Bistro, Semarang, Central Java, Indonesia

THREE-POINT PERSPECTIVE

In this section we will look at three-point perspective. Bird's-eye and worm's eye perspective are commonly used terms associated with three-point perspective, but it is important to note that technically these terms are "viewpoints", not perspective types, unless a third vanishing point is introduced.

Three-point perspective is a development of the principles of two-point perspective, but here we introduce a third vanishing point either above (worm's-eye) or below (bird's-eye) the eye level.

In diagram (A), I have drawn two stacked blocks receding towards two vanishing points (VP1 and VP2) on the eye level – this is an example of two-point perspective. In diagram (B), I have added two additional vanishing points along a vertical axis (VP3 and VP4), which alters the way the blocks are perceived, as they are now also receding to both these new vanishing points, so the size and form of the blocks seem very different. This set-up forms the basis of four-point perspective, which we will look at more later. However, if we only keep either the top or bottom vertical vanishing point, alongside VP1 and VP2, this would be a three-point perspective.

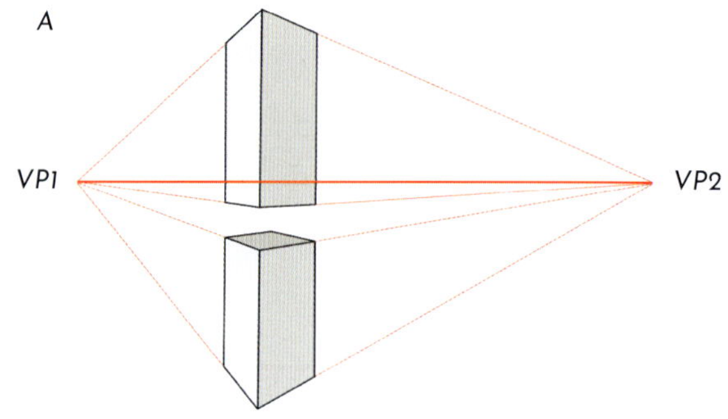

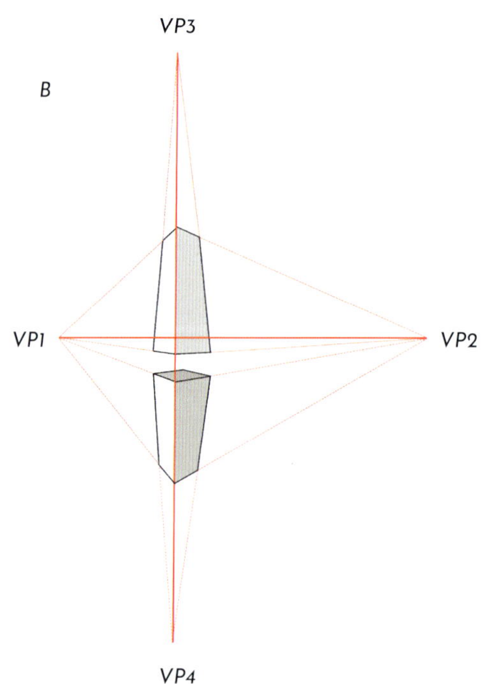

WORM'S-EYE PERSPECTIVE

Consider a two-point perspective view in a big city such as New York or Hong Kong. Look up at all the skyscrapers. Eye level is close to ground level and the tall buildings are high above us. We can establish our left and right hand vanishing points on the low eye level line but – and this is important – something new is coming into play here. As the tall buildings rise into the sky, you can see that they are also converging on a new vanishing point, and this is directly above, high in the sky. This is a third vanishing point and is a good example of what we call worm's-eye perspective, which by definition refers to a viewpoint very close to the ground, looking upward.

It is worth mentioning the third vanishing point might be off the page, so if you are doing a drawing with a third vanishing point make sure you factor this in.

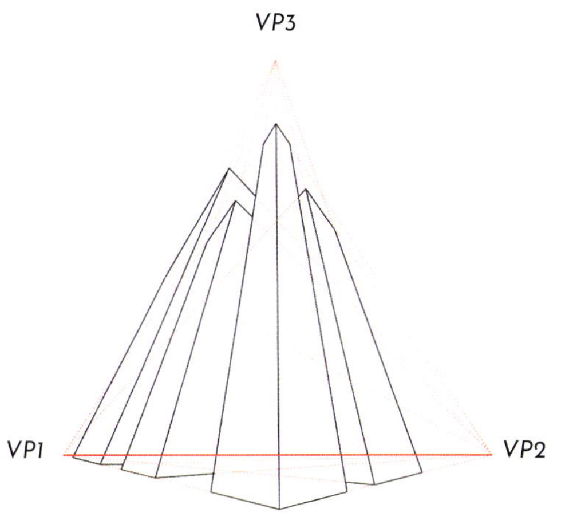

This diagram shows how building blocks, like skyscrapers in a city, recede towards a third vanishing point above.

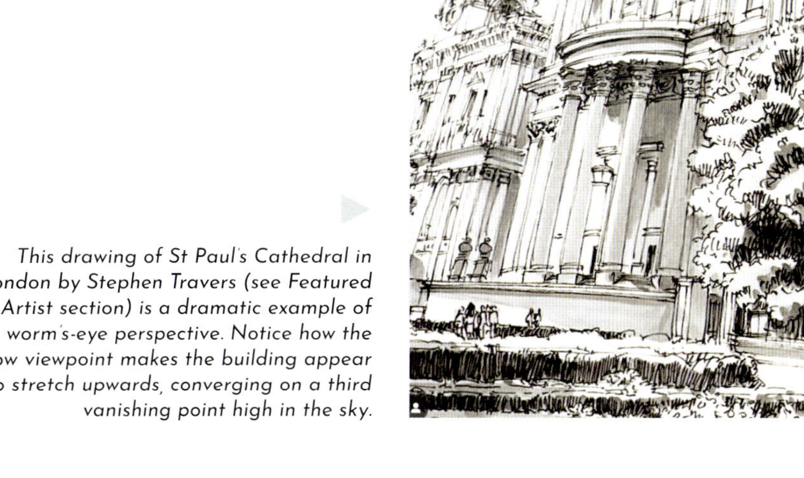

This drawing of St Paul's Cathedral in London by Stephen Travers (see Featured Artist section) is a dramatic example of worm's-eye perspective. Notice how the low viewpoint makes the building appear to stretch upwards, converging on a third vanishing point high in the sky.

You can see worm's-eye perspective in practice in this Hong Kong street view. The buildings all appear to slightly recede towards a distant third vanishing point – high above.

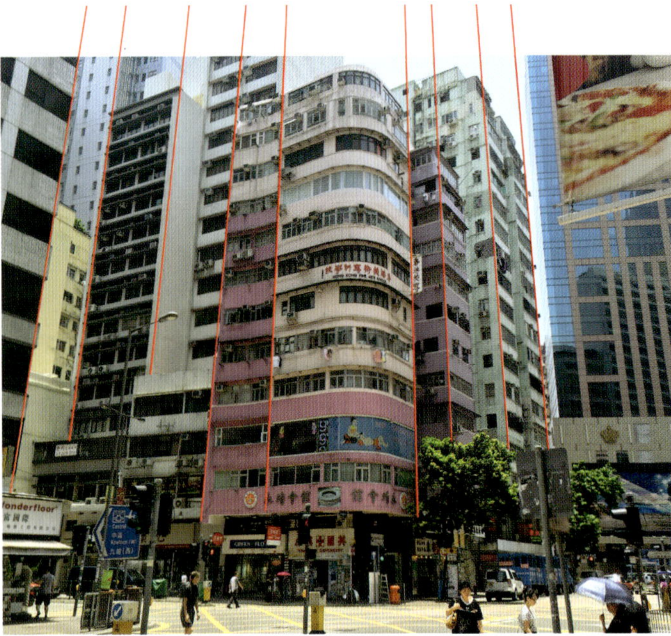

Note

In photography – as in the St Paul's Cathedral example – wide-angle lenses often create a distortion where tall buildings appear to recede dramatically towards the sky. This happens because the camera is tilted upwards, causing the vertical lines of the building to converge.

Architectural photographers often use a tilt-shift lens to correct this effect. The lens can be adjusted so it stays parallel to the building while still capturing the full height. The result is that the building looks much closer to how our eyes perceive it in real life.

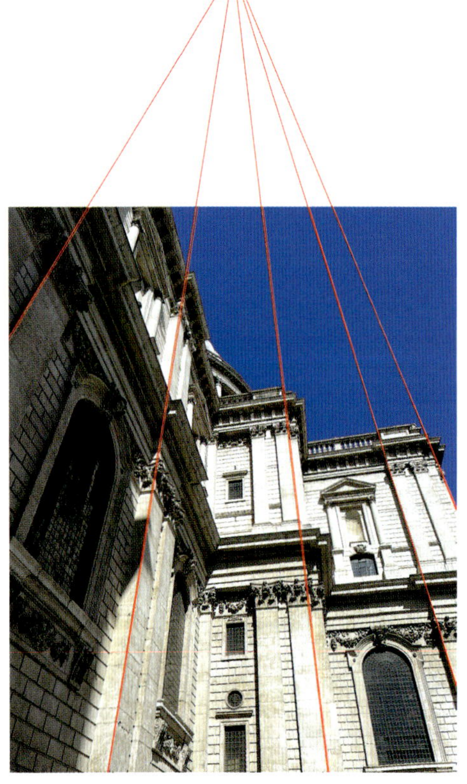

Look at this photograph of St Paul's Cathedral in London. You can see how far above the cathedral this third vanishing point sits. Because the photograph was taken with a wide-angle lens, some distortion is introduced: the sides of the building appear to lean inwards towards the sky, even though we know their façades are vertical.

BIRD'S-EYE PERSPECTIVE

The inverse of a worm's-eye view is a bird's-eye view. In its simplest form, this is also a type of three-point perspective: the eye level sits above the main elements of the scene, while the third vanishing point lies far below it. In a bird's-eye perspective, this effect is heightened because the viewpoint is positioned high above the subject – an aerial view – so we are looking sharply downward. From this height, the perspective becomes dramatic and can often appear exaggerated.

This three-point perspective diagram shows the principle of a bird's-eye viewpoint. The third vanishing point, where all the vertical lines converge, enhances the dramatic sense of height and depth.

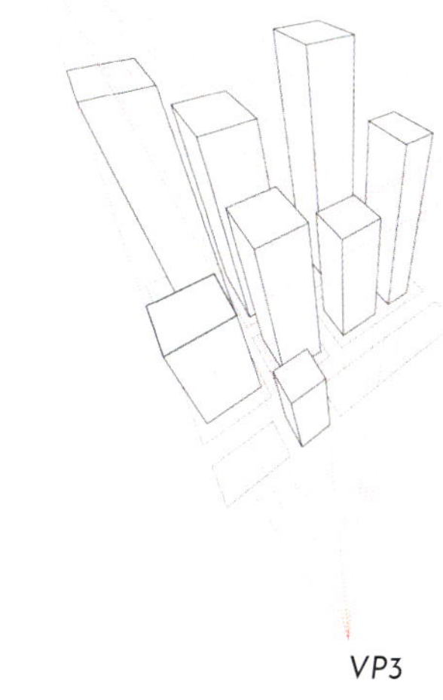

This aerial photograph of Cavendish Square in London is a good example of a bird's-eye view, taken from an aerial viewpoint. You can see that all the buildings recede to a central vanishing point far below them; however, because the photo is cropped we can't see where eye level is – but the viewpoint would suggest it is high above the square.

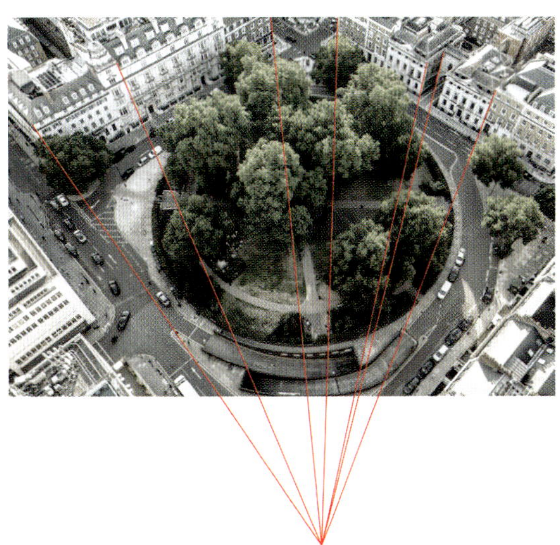

Tip

If you're drawing from a worm's-eye viewpoint, the perspective naturally creates a dramatic view! Artists who use this kind of perspective often exaggerate the receding lines towards the third vanishing point in the sky, enhancing the sense of depth and impact in the drawing.

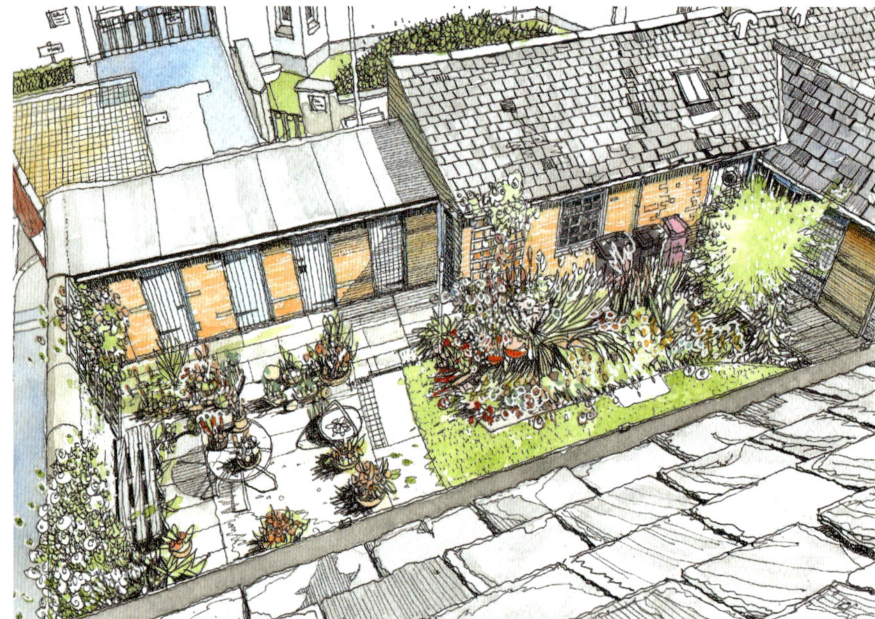

Courtyard View by Willie Watt. This sketch offers a detailed bird's-eye view of the buildings and outbuildings surrounding a courtyard garden. Although it wasn't set up using a formal perspective grid – Willie tells me he simply trusted his eye – the drawing reads unmistakably as an aerial view. The intricate textures and materiality of the façades and roofs, combined with the richly coloured garden below, create a lively and highly accomplished illustration.

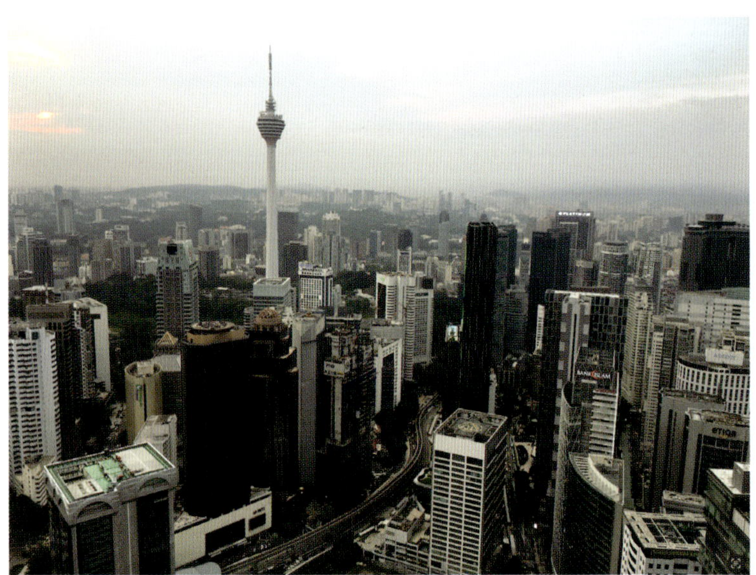

This photograph of Kuala Lumpur's cityscape also illustrates a form of bird's-eye view, though in a very subtle way. From this elevated vantage point, you can just see how the buildings taper slightly as they recede towards the ground. The vanishing point in this case lies far below the base of the buildings, which is why the perspective effect feels less dramatic than in more pronounced bird's-eye views. Even so, the image demonstrates how perspective shifts when we look down from an aerial viewpoint.

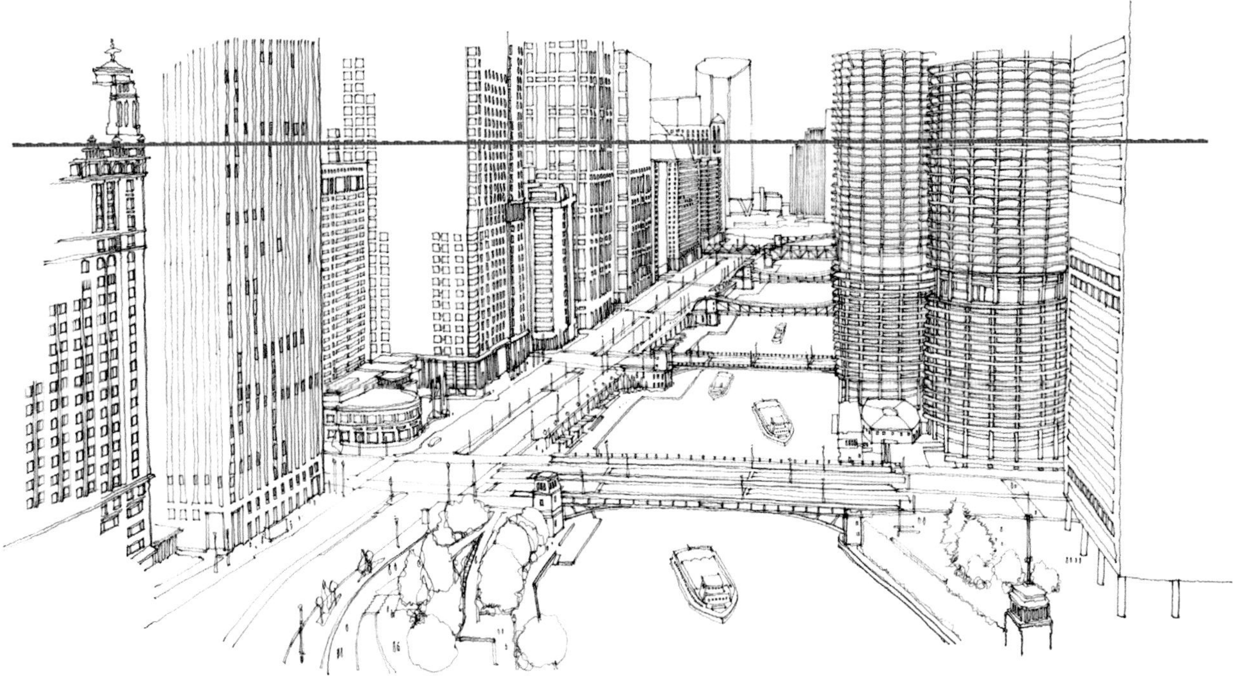

This is a commissioned illustration of the Chicago River by Vocon, an architectural and design practice with offices in New York City, Cleveland, and Chicago, USA. This is an example of a bird's-eye viewpoint which isn't a three-point perspective. It is clearly a one-point perspective, as you can see from the indicated eye-level line. Despite this being at a high level, there is no evidence of a third vanishing point in this illustration.

EXERCISE

This stylized sketch of Marina City in Chicago by architect and illustrator Carlos Almeida, demonstrates a clear bird's-eye perspective, with the vanishing point placed deep in the ground. His use of hatched shading adds depth and strength to the drawing.

As you look at this aerial-perspective drawing, try to identify the third vanishing point. Notice how the vertical lines appear to converge downward – follow them and see where they meet.

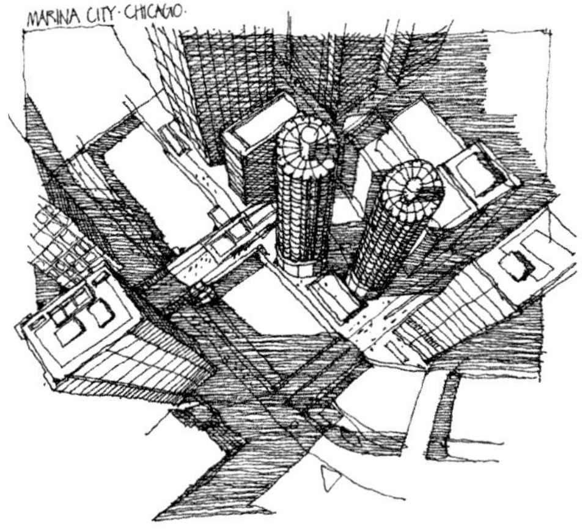

MULTI-POINT PERSPECTIVE

While it is important to understand the fundamentals of one- and two-point perspective, these systems rely on orderly, structured layouts. In real towns and cities however, streetscapes are often far more irregular and complex. They rarely follow strict patterns, which means they do not always fit neatly within these perspective constraints.

When we look at a scene that initially appears to be in one-point perspective, closer observation shows that different buildings, or groups of buildings, actually recede towards different vanishing points. Even though these vanishing points may vary, they all sit on the same eye level. This is what is known as multi-point perspective. But it is important to note, in this type of multi-point perspective there is still only one eye level, albeit with different vanishing points. Something different comes into play if we are looking at a view up or down a hill.

This is still a type of multi-point perspective but with additional vanishing points either above or below eye level (see Port Isaac photo opposite). In a quick sketch or painting, it might seem too complex to fully work out multiple vanishing points, but it's still helpful to understand the basic rules. If you're roughly following the principles of perspective, the drawing should still look realistic.

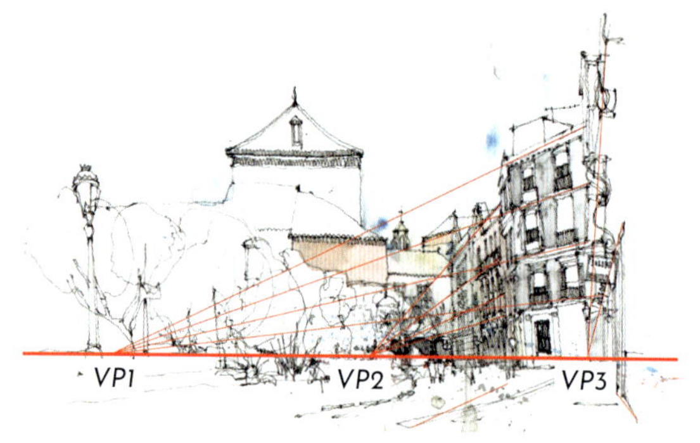

Depending on the view, there may be any number of vanishing points; if you look at this sketch drawn in Malaga, Spain, you can see three clearly identified vanishing points and they are all on the same eye level.

Tip

The most important thing to remember when faced with a complex view is to trust your eyes. First, establish whether all the vanishing points are on the same eye-level line or if there is a change in level. If there is, you may be looking at additional vanishing points above or below eye level.

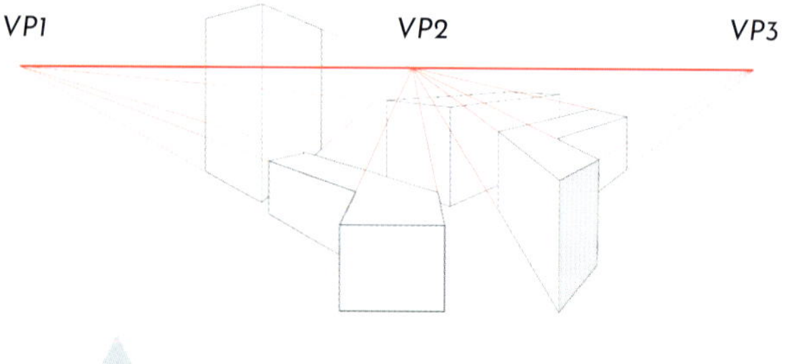

This diagram clearly illustrates how building blocks of various shapes recede towards three vanishing points all along the same eye level.

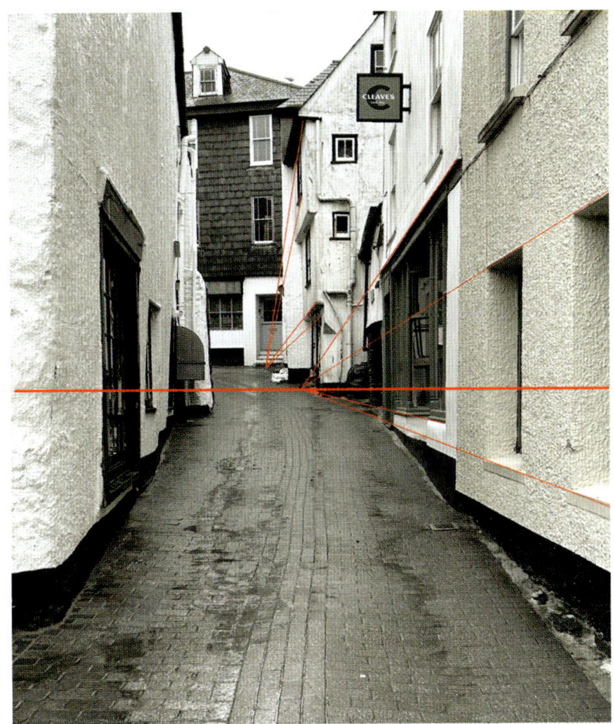

In this photograph of Port Isaac, Cornwall, UK, I have identified the eye-level line. However, because the view is on a slight incline, the more distant buildings recede towards a vanishing point just above my eye level. With historic buildings, it can be difficult to determine the exact perspective because they are often irregular. As a general guide, buildings that slope uphill away from the viewer will recede to a vanishing point above eye level, while buildings that slope downhill will recede to a vanishing point below the eye level.

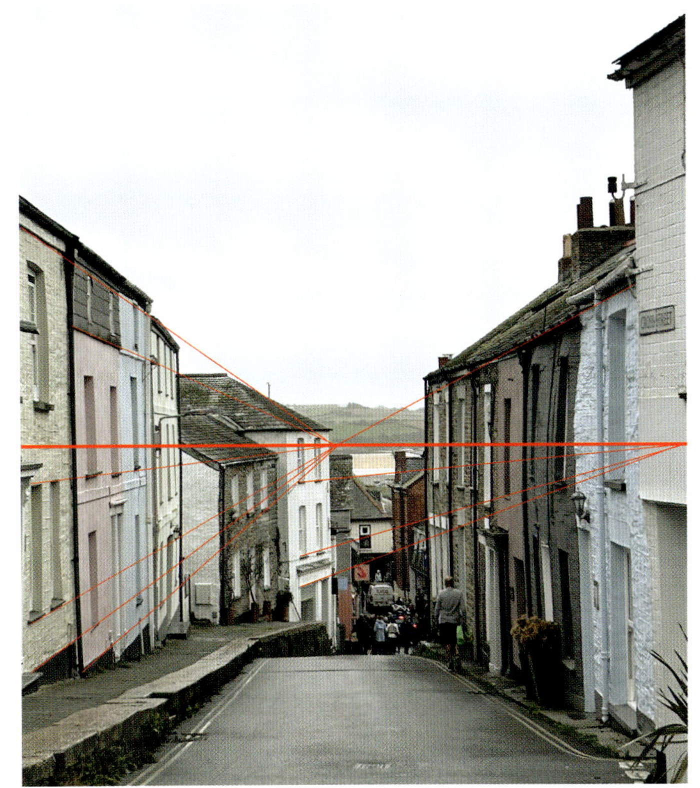

This view of a steep street in Padstow in Cornwall, UK, is more complex example of mutli-point perspective. Here I've simply picked out a couple of vanishing points on my eye-level line. In practice, if I were sketching this scene, I wouldn't worry about getting everything perfectly accurate – it is, after all, just a sketch. It is also important to say that I am looking straight ahead not down the hill, which is why there is only one eye-level line.

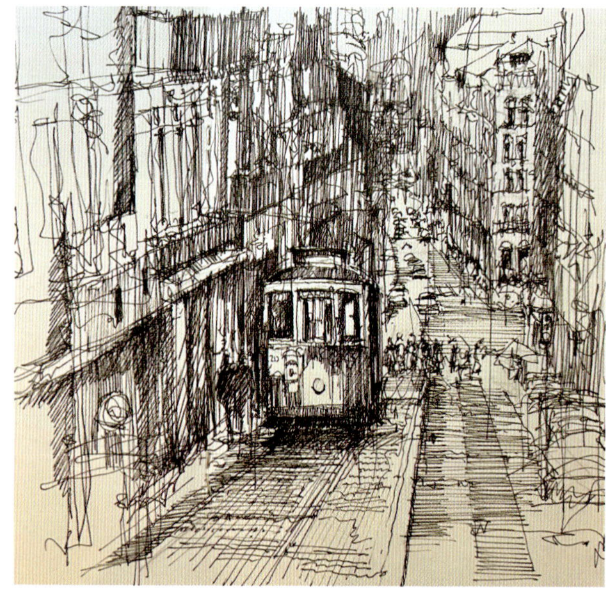

Caroline Smith captures the perspective view beautifully in this sketch of a tram in Porto, Portugal. The tram itself is on an incline, climbing towards us, while the street flattens out in the distance. Also, figures diminish in size, reinforcing a clear sense of scale. What I especially admire is Caroline's energetic use of predominantly vertical lines, which gives the drawing a fluid, dynamic quality while still being firmly grounded in accurate perspective.

EXERCISE

Look at this photograph of Telegraph Hill in San Francisco and try to work out where the vanishing points might be – it's not easy! It's not helped by the atmospheric perspective also at play here. The buildings appear to be perpendicular, so while there will be multiple vanishing points, it is a clear one-point perspective view and the vanishing points should be directly above each other.

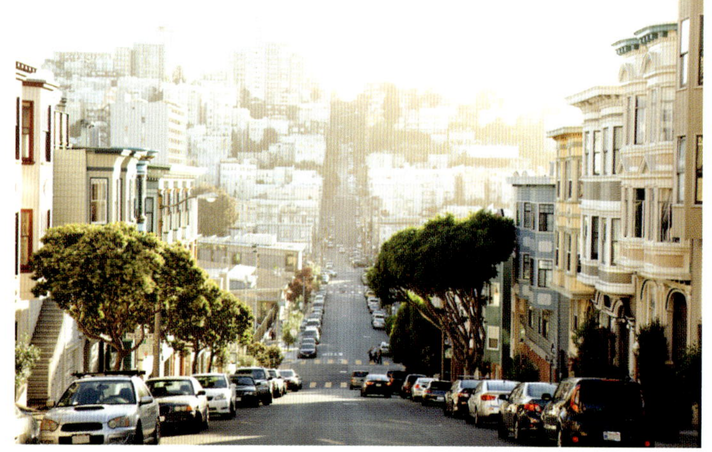

CURVILINEAR PERSPECTIVE

It's worth noting the difference between curvilinear perspective and drawing circular or elliptical buildings. While you *can* still use curvilinear perspective to draw circular buildings or objects, they are more often drawn using two-point perspective.

Curvilinear perspective creates a wide-angle, wrap-around effect. This is similar to what you see in a panoramic photograph when buildings appear to be slightly curved – especially outside our cone of vision. This is the wide-angle lens of the camera compensating for our peripheral vision. Rather than straight lines receding to the vanishing point, the lines are curved.

Curvilinear perspective is a more stylized form of perspective drawing. It is popular with some artists – especially graphic or comic artists as it has a distorted (sometimes known as "fish-eye") viewpoint. However, to achieve this wrap-around, curvilinear effect, extra vanishing points – four, sometimes five – can be required.

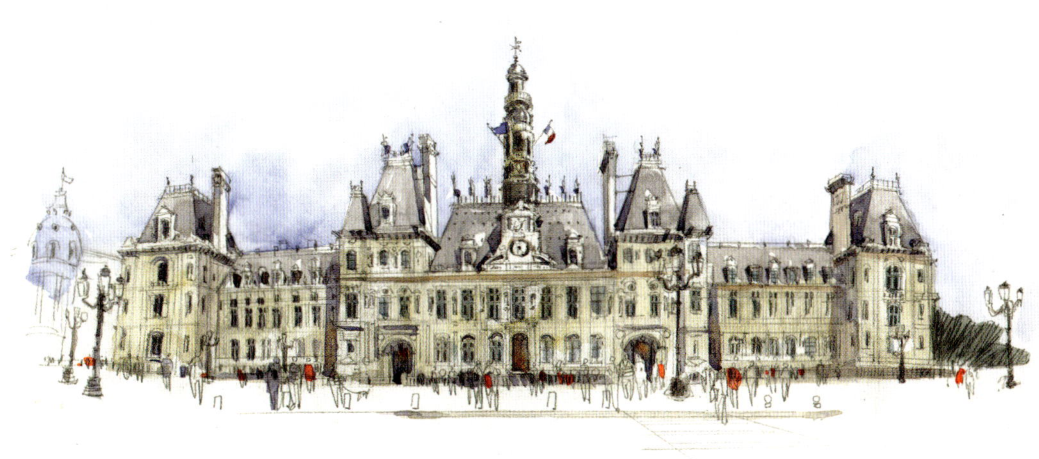

This sketch of the Hôtel de Ville in Paris by Stephanie Bower is a masterclass in subtle, curvilinear perspective. Although drawn from a direct viewpoint, both wings gently curve towards vanishing points beyond the page.

FOUR-POINT PERSPECTIVE

Four-point perspective uses four vanishing points that correlate to north, south, east, and west (see VP1–VP4 in the diagram shown here). The east and west points sit on the eye level, while the north and south points sit on a vertical axis. Where the building or object sits in this four-point perspective viewpoint will determine which vanishing points they recede to. Lines in the drawing align with these vanishing points – sometimes kept straight or exaggerated into curves. This method is often said to come closest to how the human eye naturally sees perspective.

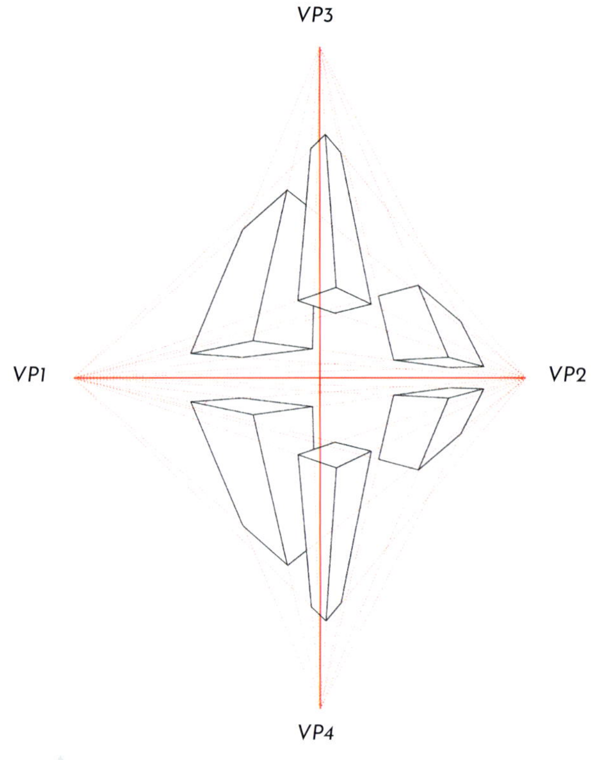

Here you can see an example of a four-point perspective grid with simple forms receding towards all four vanishing points.

FIVE-POINT PERSPECTIVE

Five-point perspective adds an extra fifth vanishing point at the centre, where the north–south and east–west axes intersect. By sketching a light grid along both axes (see diagram), I have created a base for the drawing: verticals bend towards the north and south points; horizontals bend towards the east and west. The key difference is that anything receding also goes to the central vanishing point (VP5).

This approach can be used to show curved buildings or for interior views where the outer walls appear to wrap around the central space. It's especially popular with cartoon and comic artists, as the wide-angle fish-eye effect creates a bold, immersive look that works perfectly for dramatic storytelling. This diagram demonstrates simple forms receding towards all five vanishing points. The centrally positioned vanishing point creates the curved appearance of the forms.

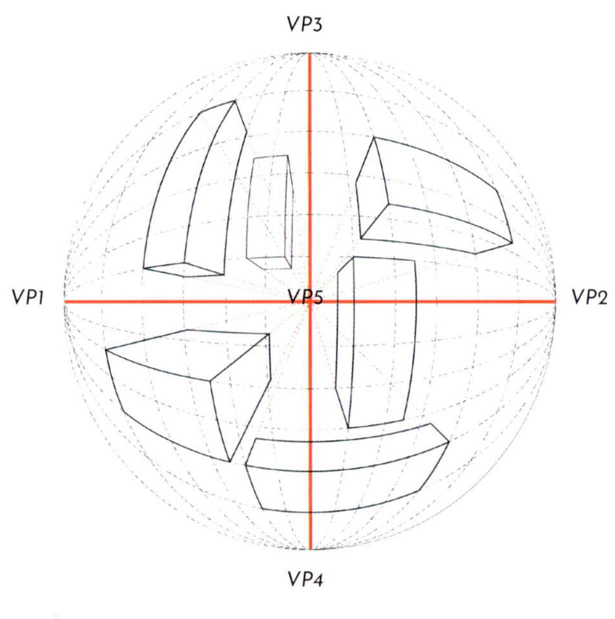

Diagram showing a central vanishing point

One of the best ambassadors of curvilinear perspective drawing is US artist and urban sketcher Paul Heaston. He regularly draws using this style of perspective, often interior views in which he exaggerates the principles of curvilinear perspective to include himself in the drawing. In this sketch of the Tattered Cover Bookstore in Denver, USA, Paul has provided an overlaid grid over his original sketch. The way he has positioned his sketchbook at the south axis and how this draws himself into the sketch is a bold move.

The overlay demonstrates this five-point perspective view. The red lines recede to the horizontal vanishing points (east and west) on the eye level. The green lines recede to the bottom vanishing point (south), and the blue lines to the central vanishing point. Although Paul has not shown any lines receding to the top vanishing point, the five-point perspective view is implied.

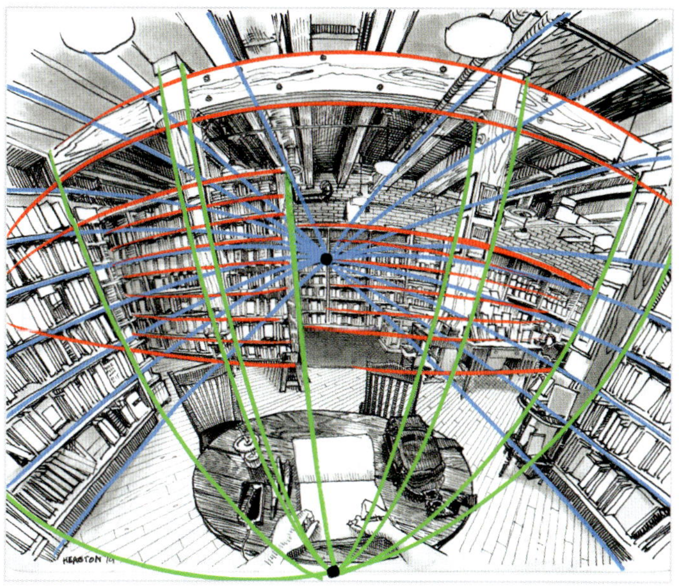

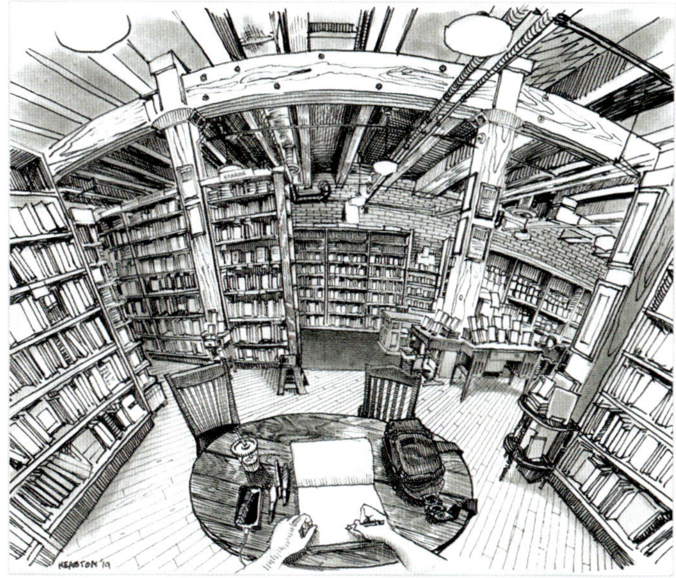

FEATURED ARTIST

Paul Heaston

Denver, USA

Over the years I've narrowed down my materials to a few favourites.

I carry a Sailor Fude de Mannen fountain pen and a water brush filled with diluted grey waterproof ink everywhere I go. The Sailor Fude pen has a bent nib that allows me to vary the line weight just by changing the angle at which the nib comes in contact with the paper. I can get a very bold line like a marker and also a very thin line suitable for fine detail or hatching without having to reach for another pen. The water brush lays down ink washes that help me add value quickly as a supplement to my hatching and cross-hatching. Because I use a waterproof ink for the washes, I can layer them to create even darker values without worrying about lifting the previous layers.

I've also become very interested in exploring unconventional perspective approaches to the spaces I sketch, particularly interiors. I love the challenge of attempting to capture my whole field of view within the sketchbook page, including my hands and the sketchbook itself! This leads to a "fish-eye" effect that seems to show the viewer my environment as though they are looking directly out of my eyes, so I call them "POV" or "point of view" sketches.

This sketch by Paul Heaston is a perspective masterpiece, drawing the viewer into a quiet domestic scene that includes the artist himself and his wife, Linda. Though it appears familiar and suburban at first glance, it reveals extraordinary technical skill, enhanced by Paul's effective use of colour.

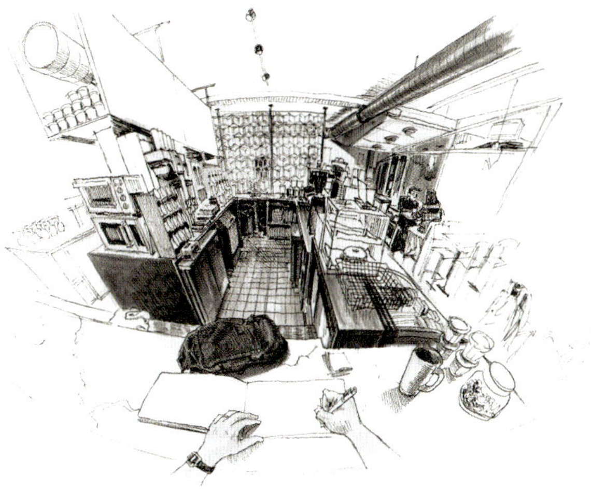

The Bardo Coffee House, Wheatridge, Denver, USA

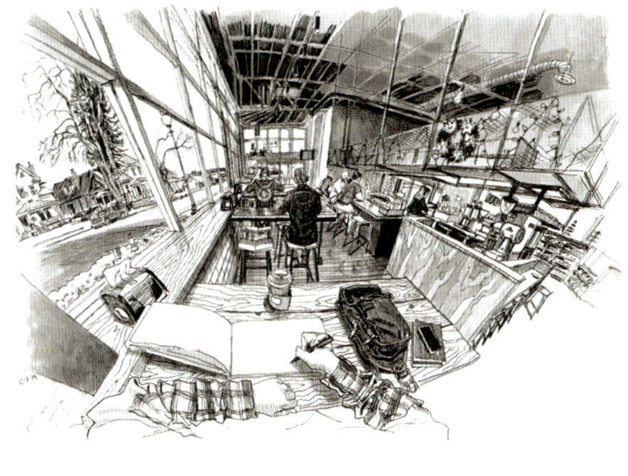

Leevers, Locavore, Natural and Organic Foodstore, Denver, USA

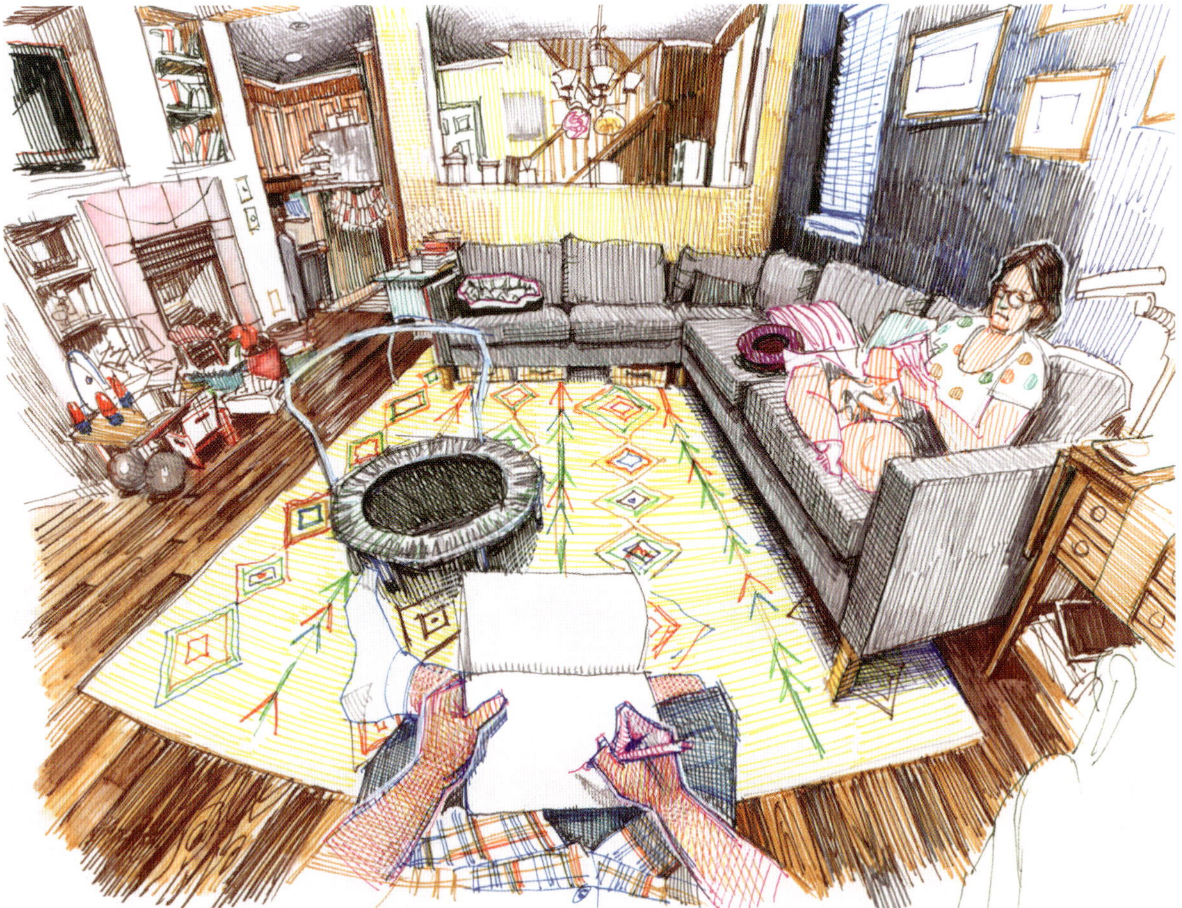

THE CITY IS A STAGE

Think of your drawing as a stage set! This drawing was part of a workshop proposal for the Urban Sketchers Symposium in Buenos Aires, Argentina. My idea was to explore how to simplify complex cityscapes - because in a big, complicated city, the hardest question is often: where do you even begin sketching?

The technique I proposed was borrowed from the simple principles of stage-set design. The city view becomes a series of theatrical layers:

- Down-stage (foreground)
- Centre-stage/performance area (main focal point)
- Up-stage (background).

I also wanted to bring in the principles of stage-left and stage-right by studying composition and referring to the rule of thirds in how key points of interest are positioned. By treating the complex city view as a series of layers, background noise and urban clutter are quietened, making the urban environment easier to understand and sketch.

I've adapted this drawing of Plaza de Mayo in Buenos Aires to demonstrate my approach, overlaying a rule-of-thirds grid to help illustrate the composition. The foreground - or down-stage - urban landscape is drawn with the heaviest line weight. In the middle ground, or centre-stage, are my main focal points: the Cabildo Museum, slightly off-centre, and the Metropolitan Cathedral to stage-left (which, from our perspective as viewers, appears on the right).

I've intentionally chosen not to emphasize the central monument, as my interest lies more in the surrounding buildings. The monument's perfectly central position doesn't lend itself to a dynamic composition. I should note that this drawing was done from Google Street View, as I haven't yet had the opportunity to visit Buenos Aires in person. Were I drawing on location, I would have sought a viewpoint where the monument wasn't so centrally placed.

That said, I've done my best to ensure the line weight of the monument doesn't distract from the main focus of the sketch. For the background - or up-stage - architecture, I've used lighter lines to convey depth and distance.

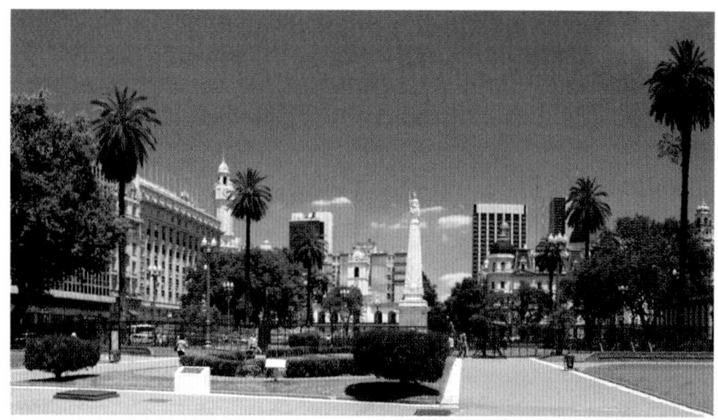

Plaza de Mayo, Buenos Aires, Argentina

Plaza de Mayo, Buenos Aires

Note: stage-left is on our right when we're in the audience.

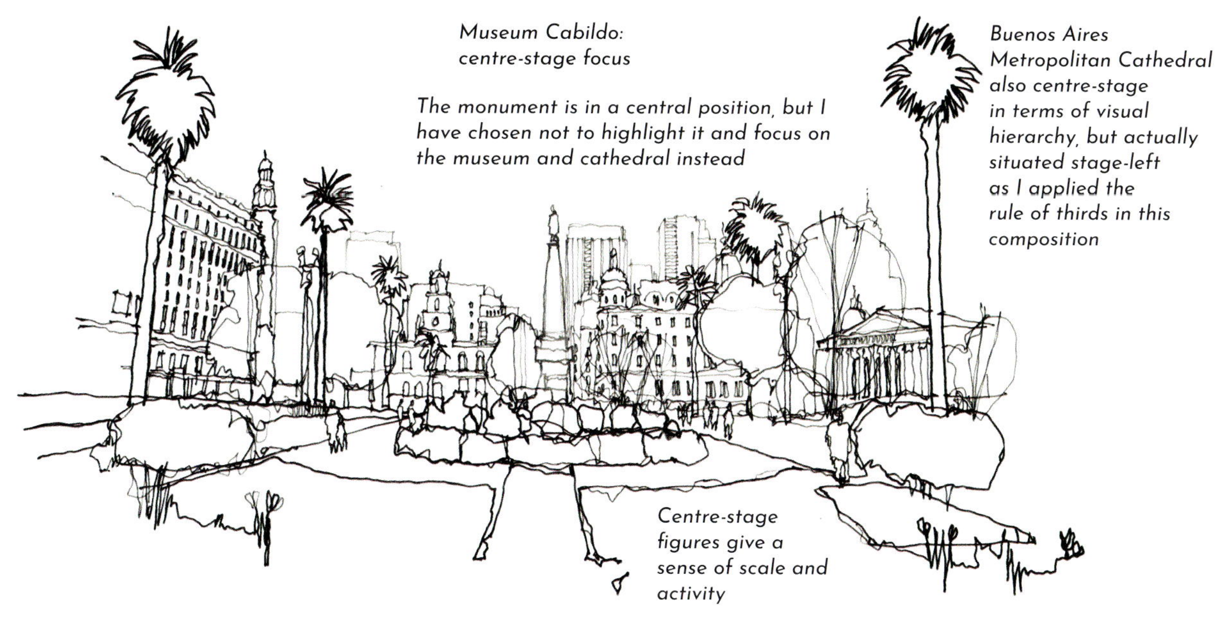

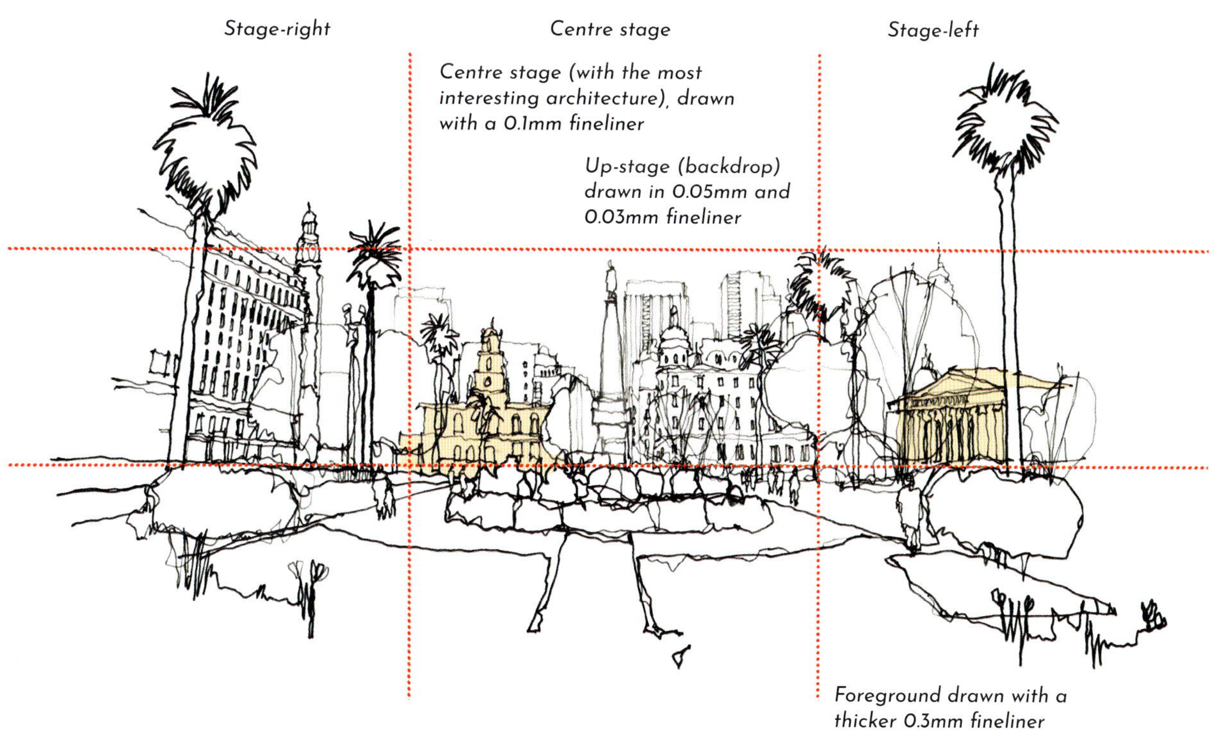

SHADOWS

Shadows occur when an object or structure blocks a light source. Their form is shaped by the object's geometry and follows the rules of perspective, whether the light source is natural or artificial.

Before looking at shadows in perspective, it is important to understand how they behave under normal conditions. The length and direction of a shadow is determined by the sun's position in the sky. When the sun is low and near the horizon, shadows are long and extended. As the sun rises higher, shadows become shorter and more compact. At midday, when the sun is overhead, shadows are at their minimum and may almost disappear.

We also have evening light. Again, this is when the sun drops nearer the horizon, and there are lengthened shadows because the angle of the sun's rays are low. It's also interesting to note at this time even small objects can cast long shadows.

And, of course, without sunlight, there are no shadows at all!

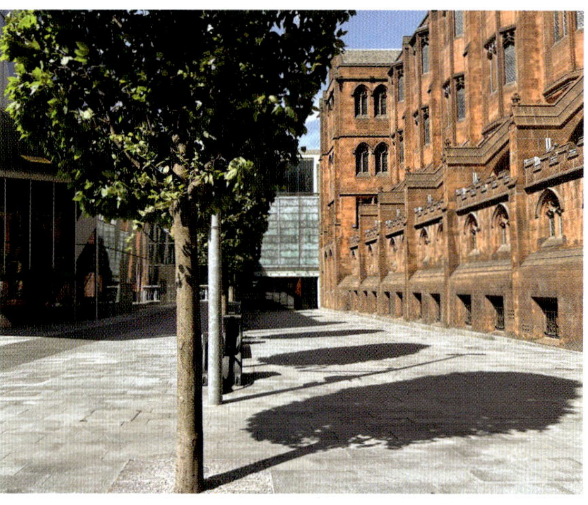

Notice the long shadows cast by the trees and the darker, more pronounced shadow of the building to their left. From these clues, we can deduce that the light source – the sun – is positioned high in the sky, towards the left side of the photograph.

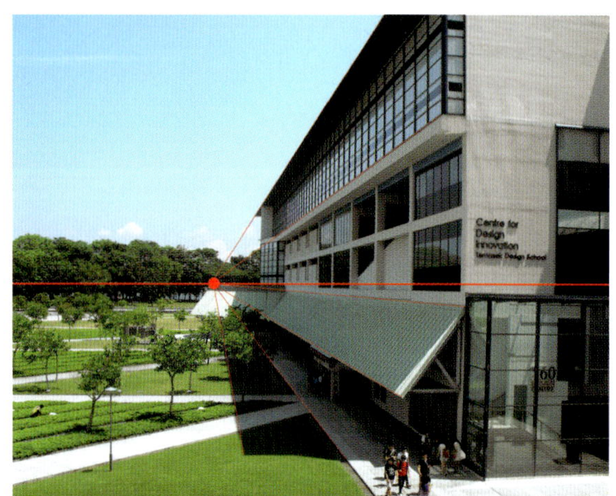

This photograph of Temasek Polytechnic in Singapore, taken in bright sunlight, clearly shows how the shadow cast by the ground-floor canopy recedes towards the same vanishing point as the building itself.

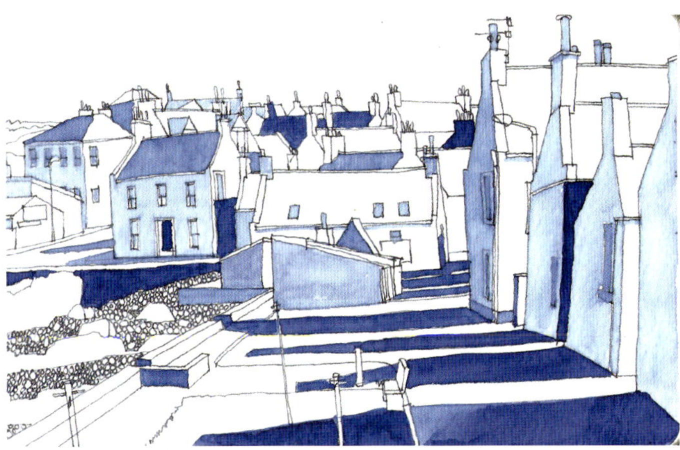

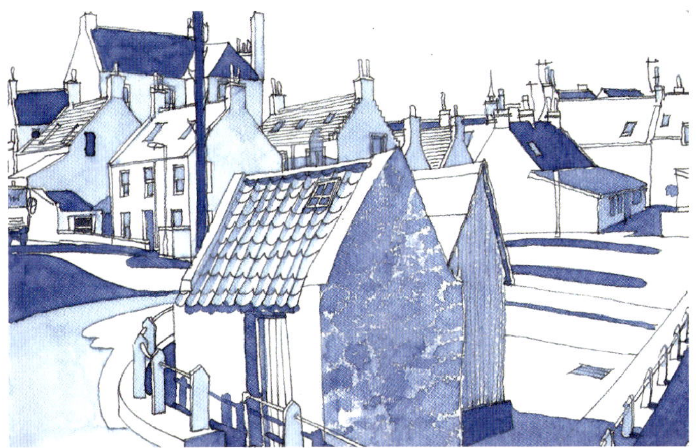

These sketches, with their strong shadows, are elegantly rendered – Willie Watt's distinctive style is immediately recognizable. These two sketches, part of what I would describe as his trademark "blue series", clearly show how dramatically cast shadows fall across buildings.

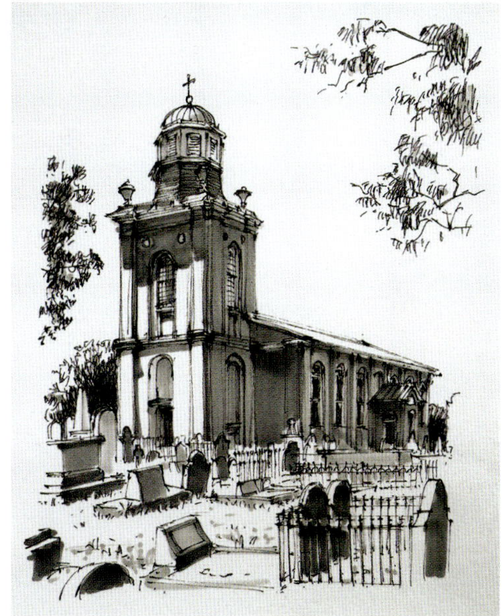

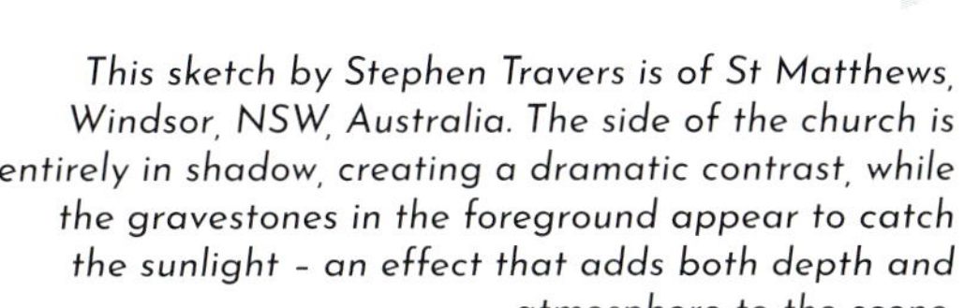

This sketch by Stephen Travers is of St Matthews, Windsor, NSW, Australia. The side of the church is entirely in shadow, creating a dramatic contrast, while the gravestones in the foreground appear to catch the sunlight – an effect that adds both depth and atmosphere to the scene.

HOW TO DRAW SHADOWS

1. Draw your eye level.

2. Next, draw a cube in two-point perspective that recedes to two vanishing points (VP1 and VP2), but keep the cube beneath the eye level for this example.

3. You then need to choose your light source location – this doesn't have any relationship to your existing vanishing points.

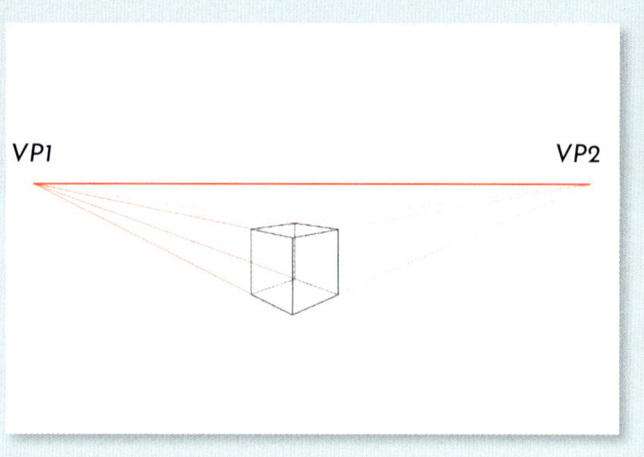

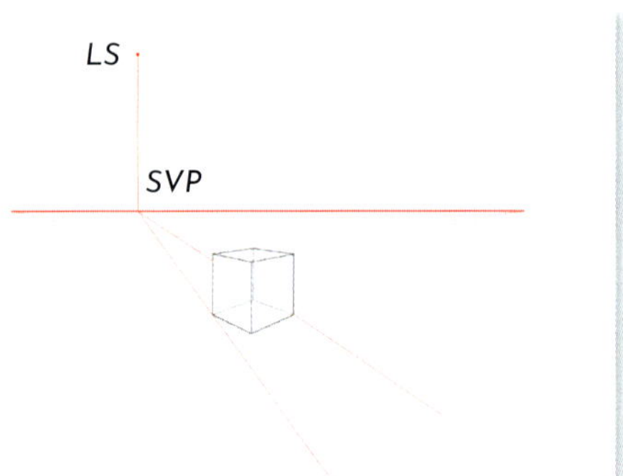

4. Draw a vertical line from your light source (LS) to the eye level – this is your shadow vanishing point (SVP) and will be where your shadows will recede to; draw two lines from the bottom left- and bottom right-hand side of the cube back to this vanishing point and extend forward to the edge of the page.

5. Now draw a line from the light source to each *visible* corner of the cube; you could label them A, B, and C. Extend these lines as they are going to define where the shadow is located beyond the cube. The left-hand and right-hand lines (A and C) should intersect with lines drawn from those same corners back to SVP, which is the vanishing point that controls the direction of shadows on the ground plane.

6. To locate the final shadow edge, draw lines from VP1 and VP2 that pass through the intersection points created in step 5 (where A and C meet the SVP lines). These new lines define the final angles/directions of the cast shadow.

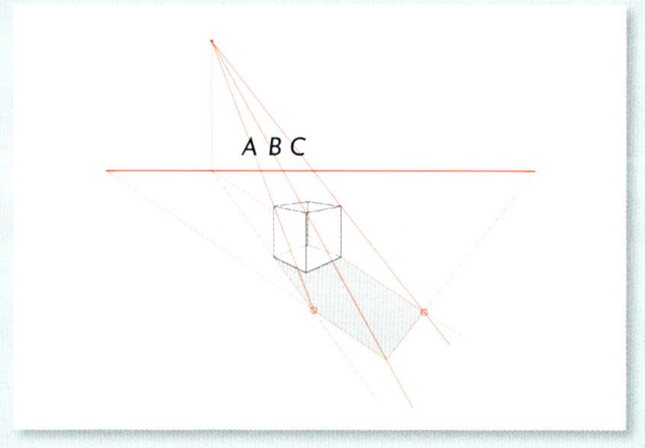

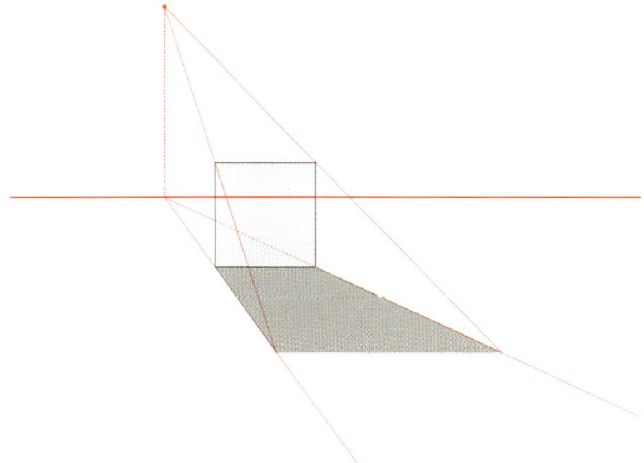
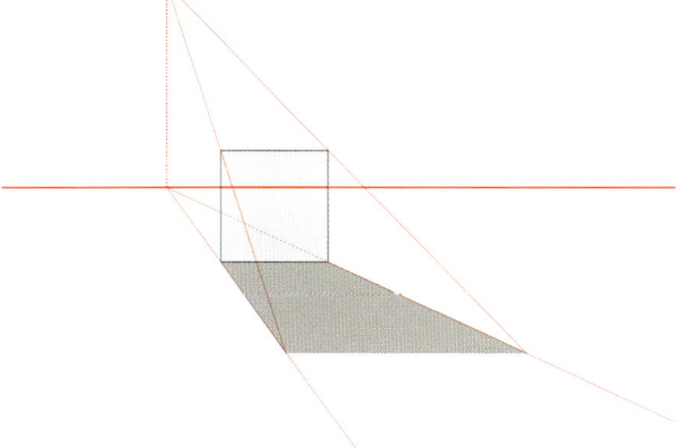

This diagram shows a vertical flat plane casting a shadow in one-point perspective. It uses the same method as the earlier two-point example but simplified. The shadow is formed where lines from the light source meet the top left and right corners of the plane, and where those lines intersect with the eye-level vanishing-point lines.

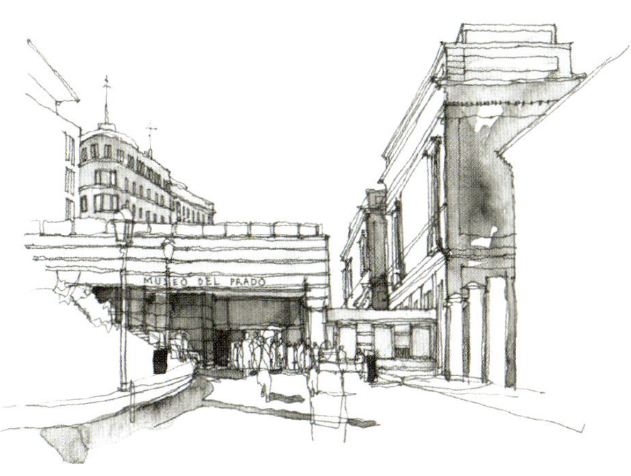

Drawn outside the Prado Museum in Madrid, Spain, in summer sunshine, this sketch uses a deliberately monotone palette. I was particularly interested in the shadows – a prompt to remember the importance of shadows cast by street lights, figures, and other urban elements.

People Cast Shadows Too!

The buildings are in sunlight, but check out the extended shadows cast by the people in this photograph looking down on the Grassmarket in Edinburgh, UK.

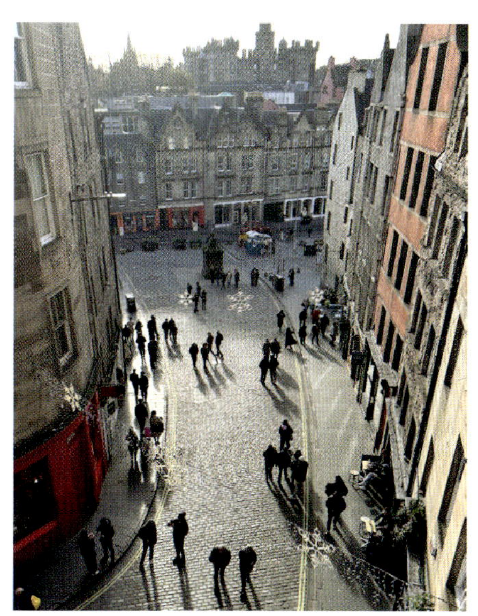

Tip

As a reminder: if the light source is higher, the shadows appear shorter; and if the light source is lower the shadows become longer and more extended.

ADVANCED SHADOWS

This is quite a complex example, but follow along if you would like to understand how to project more complex shadows.

1. Draw three buildings: two with pitched roofs and one with a flat roof. Position them below the eye-level line that they recede towards their respective vanishing points. Draw light projection lines from the buildings to these vanishing points:

- The left-hand building recedes to VP2 and VP3.
- The middle building recedes to VP1 and VP3.
- The right-hand building recedes to VP2 and VP4.

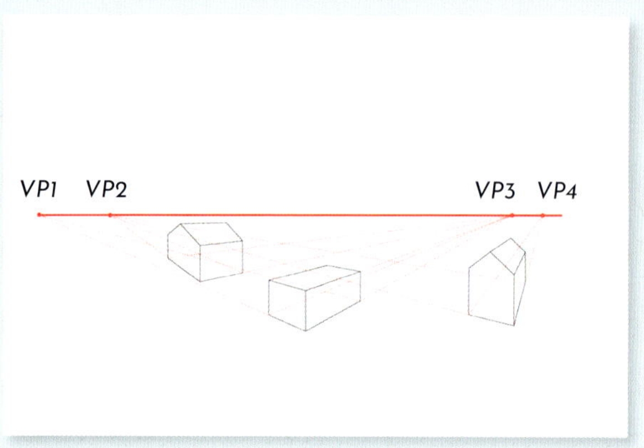

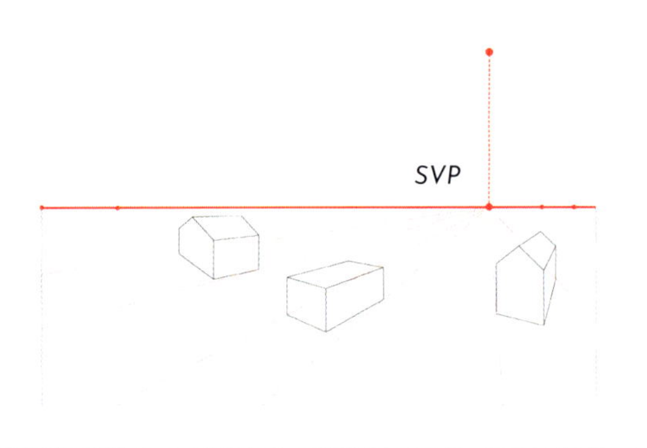

2. Next, position the light source, representing the sun. From this point, drop a vertical line down to the eye-level line. This is the shadow vanishing point (SVP). Now draw faint projection lines from the left- and right-hand corners of each building back to the shadow vanishing point. These projection lines will establish the direction in which the shadows will fall relative to the light source. You can also extend these lines beyond the buildings to a notional edge of the ground plane as this will inform where shadows are projected.

Here, I have removed the projection lines back to the vanishing points (VP1–VP4). However, if you are drawing lightly in pencil, you may prefer to leave these lines visible.

3. The middle building is the simplest to find the shadows. From the light source, draw lines through each of its three visible top corners and extend them beyond the building's outline. Label these C1, C2, and C3. If drawn correctly, C1 and C3 will intersect with the projection lines back to the shadow vanishing point.

Where C1 intersects the line from the SVP label this point P1 (P for Plan). Next, locate where the line drawn through C3 intersects the line returning to the shadow vanishing point. Label this intersection P2. From P2, project a new line back to the original vanishing point (VP3). Extend this line until it meets the line which has been extended from C2. Label this intersection P3, then draw a line from P3 back to VP1. If the geometry is correct, this line should intersect PL1. The line drawn from P1 back to the SVP forms the left-hand edge of the shadow. Finally, complete the shadow by drawing a line from P3 towards the shadow vanishing point, connecting it to the right-hand corner of the building.

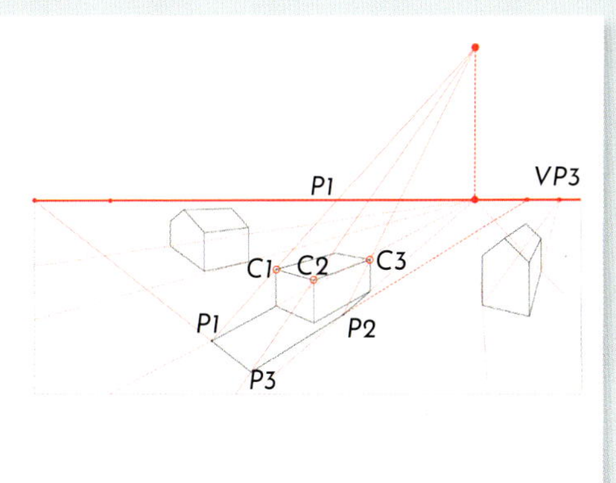

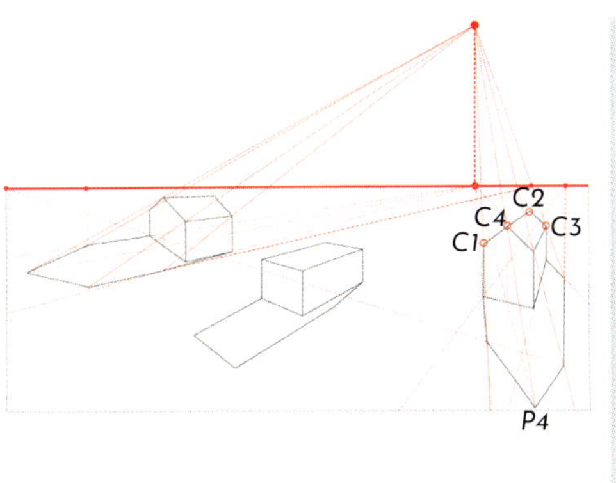

4. This same method applies to the other buildings, with the additional shadow from the pitched roof to consider. This time, use four intersection points – C1, C2, C3, and C4 – where C4 represents the apex of the pitched roof. These shadows are constructed in the same way as for the flat-roofed building. From the apex of the gable, drop a vertical line to the ground plane. From this intersection, project a line back towards the shadow vanishing point, extending it beyond the building. The point where P4 intersects this line determines the shadow angles of the pitched roof.

5. This diagram shows the completed buildings and their shadows.

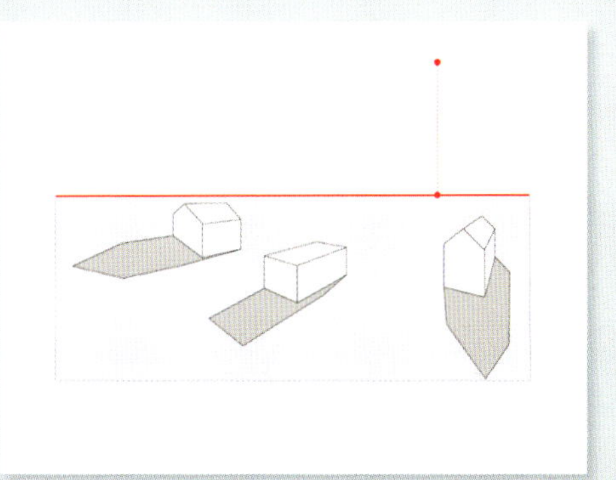

FEATURED ARTIST

Willie Watt

Dundee, Scotland

I am a conservation architect who studies Scotland's vernacular buildings – how they are built, and how they sit together in farms, villages, and around fishing harbours, forming each place's unique townscape.

Many of my paintings are created using ballpoint or fountain pens, combined with abstracted acrylic ink washes. These often use duotones, focusing on cool blues to form shadows that speak of winter alongside a hygge-like contrasting and very limited use of spot colour providing warmth via lit windows, fire-clay chimney pots, or pan-tiled roofs.

My focus on shadows pulls out three-dimensional forms within each view and often suggests buildings that sit just "off stage" but still cast shadows across the scene to emphasize that my subjects are set within a wider place. These shadows emphasize perspective by creating abstracted planes, which in turn reveal the shape and distortion of surfaces and the ground around the buildings. I often frame the gables of buildings by contrasting these with spaces around them that are in shadow, or by bookending caught glimpses of landscapes beyond.

I use acrylic ink because of its clean and transparent appearance, applying a series of washes using the same mix to achieve tonal variation.

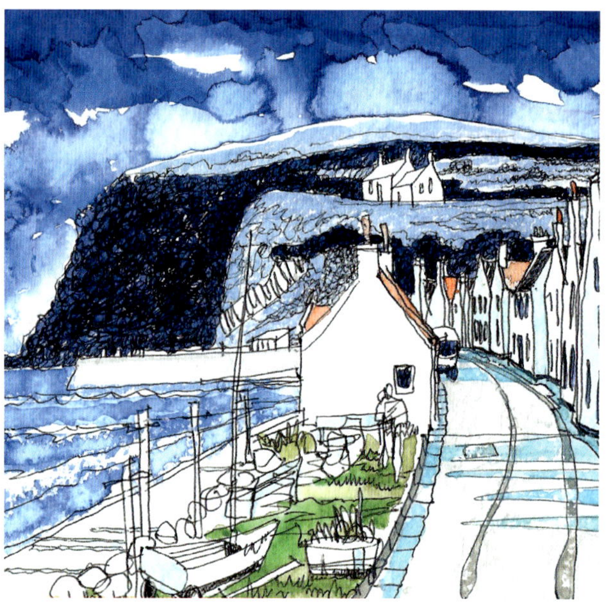

Pennan Harbour, Scotland

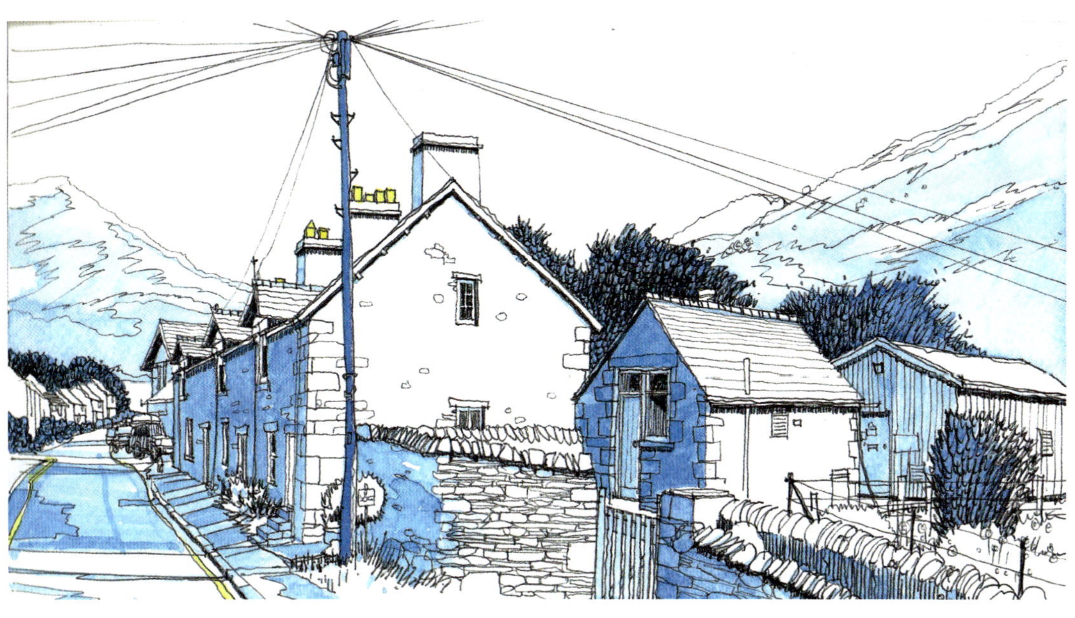

Glenridding, Ullswater, Cumbria, England

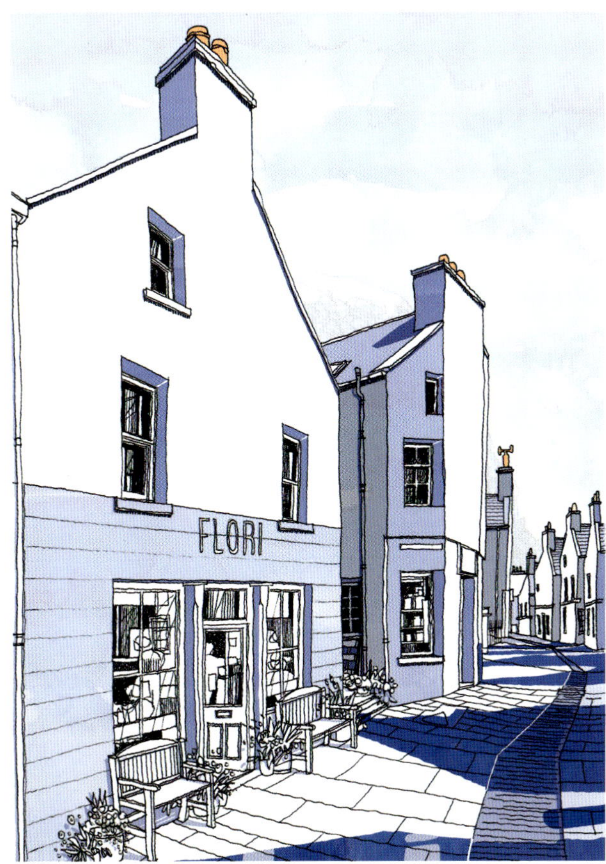

Flori, Dundas Street, Stromness, Orkney, Scotland

REFLECTIONS

When learning about shadows in perspective, it's also worth thinking about reflections as they work in a similar way. A reflection is essentially a mirror image across the reflective plane. Reflections in perspective follow a similar pattern to shadows but rather than projecting from a light source, they project across a reflective surface (such as water or glass).

If we focus on reflections in calm water, they will align with the existing vanishing points, with lines converging just as they do on the eye level. However, because vanishing points sit at eye level while the waterline is below it, the reflected elements receding towards the vanishing points may appear to taper more steeply, creating a slightly distorted but familiar perspective effect.

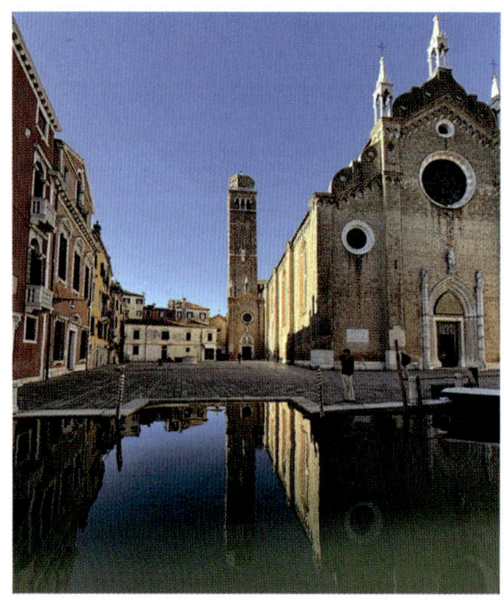

This photograph of the Basilica of Santa Maria Gloriosa dei Frari in Venice, shows how still water captures the church's reflection, receding to the same vanishing point as the physical church above. Photo by Lois Blackwell.

A

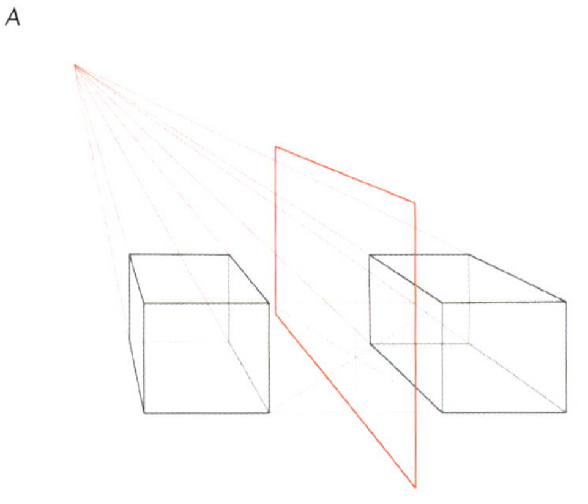

B

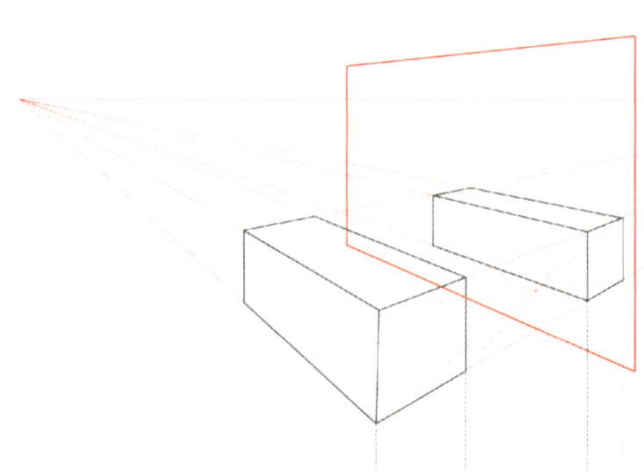

Here are two simple examples of mirrored reflections. Diagram A shows a reflection in one-point perspective, while Diagram B illustrates a two-point perspective. In both cases, the mirror line acts as the midpoint, so the reflected depth is equal on either side.

LOCATING THE REFLECTION

In this image of Amsterdam, the reflections in the water recede towards the same vanishing points as the buildings above. However, the viewer's eye level sits above the waterline, not at the point where the buildings meet the canal. Because of this, the reflected lines in the water slope slightly more steeply towards the vanishing points than the physical building above. .

> *Tip*
>
> What's above the horizon appears below the horizon in reverse, receding towards the same vanishing points.

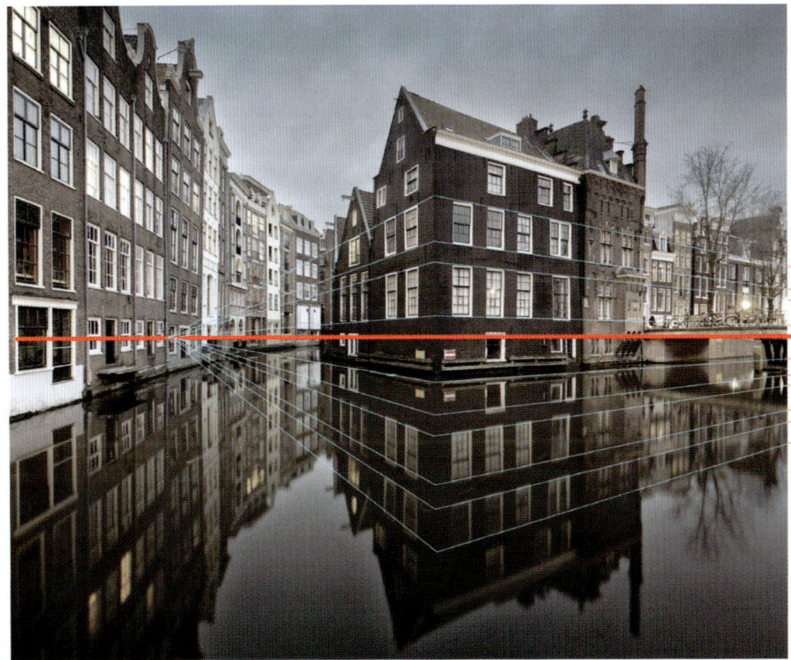

Oudezijds Voorburgwal canal, Amsterdam, Netherlands

ROOFTOPS

When drawing rooftops in perspective you can think of buildings as stacked boxes, with roofs simply forming their top planes. Rooftops recede towards the same vanishing points as the rest of the building. From higher viewpoints, such as in an aerial view, roofs appear as foreshortened rectangles converging towards the vanishing points, with their visibility depending on their position relative to the eye level.

In a city scene, all parallel roof edges should align with the same vanishing points (whether one-point, two-point, or three-point perspective, depending on the view). Finally, remember that rooftops aren't uniform. Cities are full of "rooftop clutter": chimneys, stairwell blocks, water tanks, skylights, air-conditioning units, and more. Adding these smaller box forms (also in perspective) can make the sketch livelier!

Tip

In this overlay of Karen's sketch below, I have simplified the flat roofs, which recede towards a vanishing point on the eye-level line. This can be useful when tackling a complex rooftop view or busy skyline.

Keep in mind that the lower a rooftop is relative to eye level, the more of its surface will be visible. In many cases, as these examples show, there may be little "true" perspective at rooftop level; depth is often suggested instead through atmospheric perspective techniques.

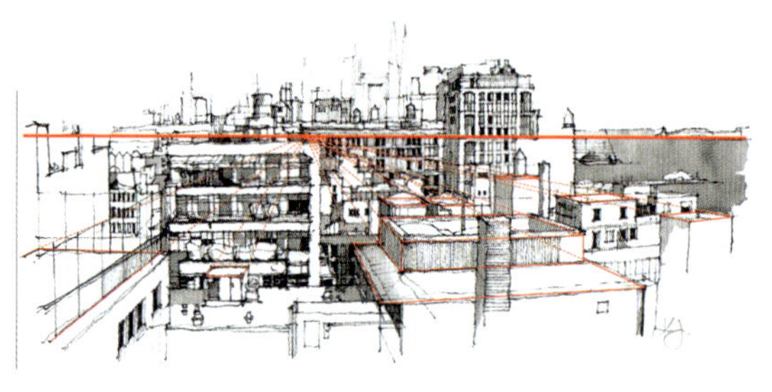

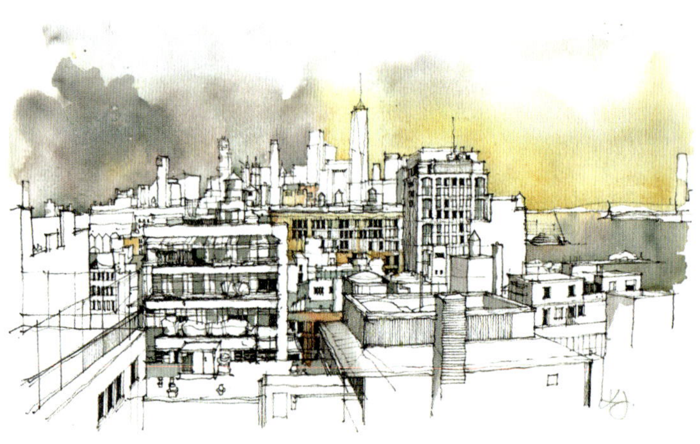

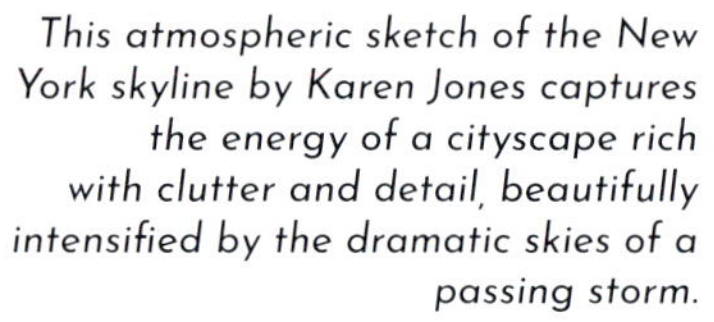

This atmospheric sketch of the New York skyline by Karen Jones captures the energy of a cityscape rich with clutter and detail, beautifully intensified by the dramatic skies of a passing storm.

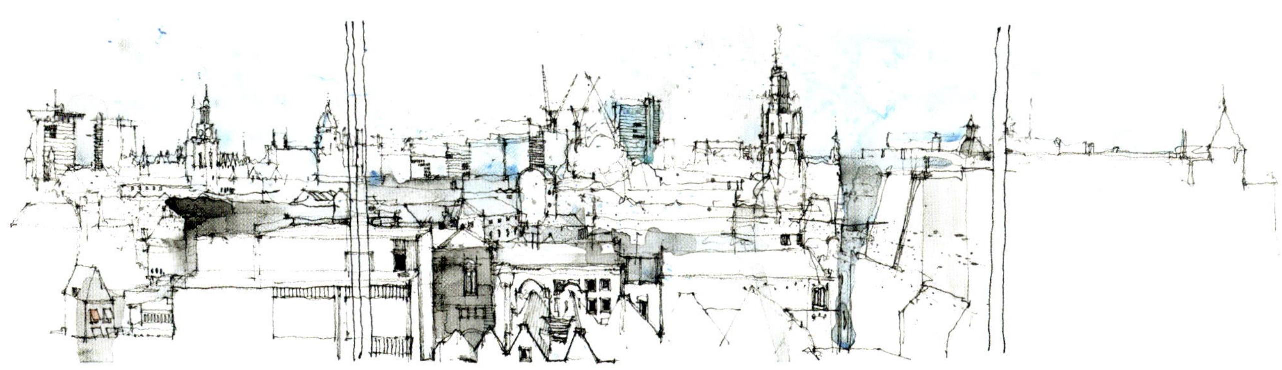

Sometimes, when sketching a busy cityscape, rooftops can appear chaotic and without any obvious pattern, as in this Amsterdam sketch. In such cases, there may be no clear perspective lines to anchor the view, yet a sense of depth is still achieved through atmospheric perspective. Details fade as they recede into the distance, while key features, such as distinctive buildings – in this case, church towers on the skyline – provide strong focal points that hold the composition together.

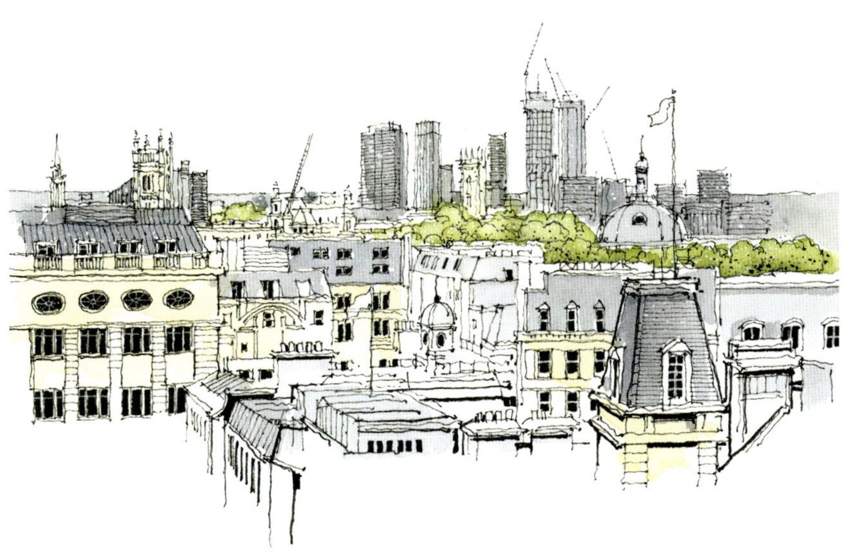

This sketch of London, UK, by Phil Dean feels very structured, with a concentration of detail in the foreground. A high eye level – set a few storeys above ground – is clearly established, with buildings and rooftops receding towards a vanishing point on the left. The distant skyline is rendered more loosely, shown in elevation rather than perspective. Though lacking recession, the reduced level of detail also effectively conveys atmospheric perspective.

FEATURED ARTIST

Phil Dean

London, UK

Perspective is the thing that sketchers often worry most about. Technical talk about one-point, two-point, and multiple vanishing points can be daunting, but in practice it's not difficult to get the hang of it.

My advice is always to treat perspective with respect but don't let it rule your drawings. I firmly believe that wonky or "wrong" perspective adds character to a sketch, making it feel more real than a technically accurate perspective drawing.

It's good to have a basic knowledge of the principles of perspective so that you can have the freedom to really experiment with it. Everything you draw will follow certain laws of nature and it's only when you truly start observing that you see the principles at play. Once you start looking at perspective lines, it's impossible to unsee them.

My drawings don't follow the traditional, technically accurate rules of perspective, but I prefer a looser interpretation of the scene and I'm happy to forgive myself for any inaccuracies. There is no such thing as a mistake in a drawing!

With a scene that has complex perspective, I begin a sketch in the middle and select an angle to anchor the scene. Using a pen or pencil, I follow the angle in the drawing and transfer that angle to the page to "set" the perspective.

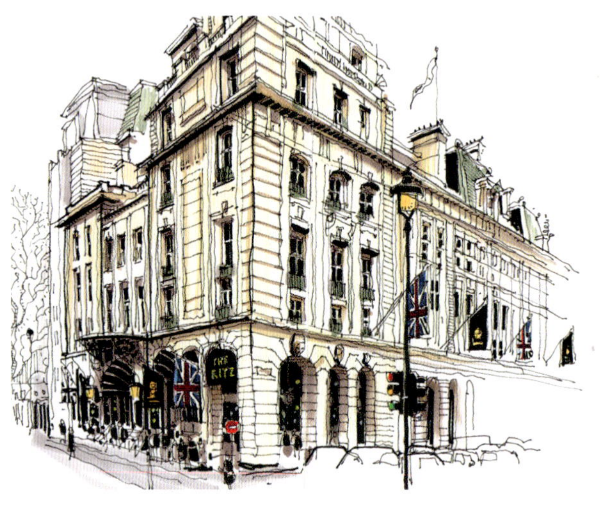

The Ritz Hotel, Piccadilly , London, UK

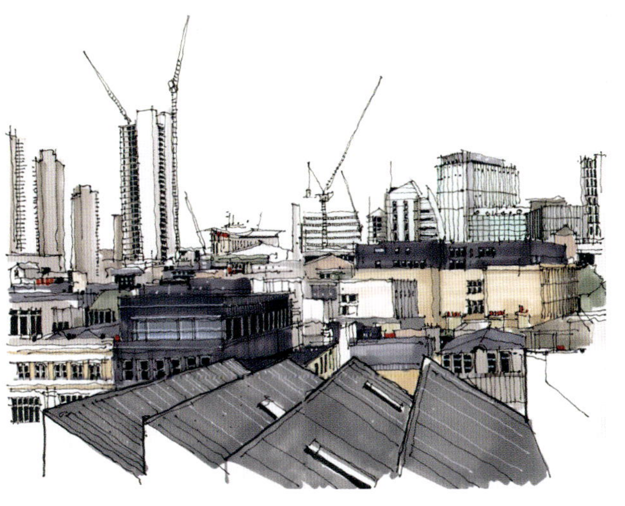

Shoreditch, London, UK

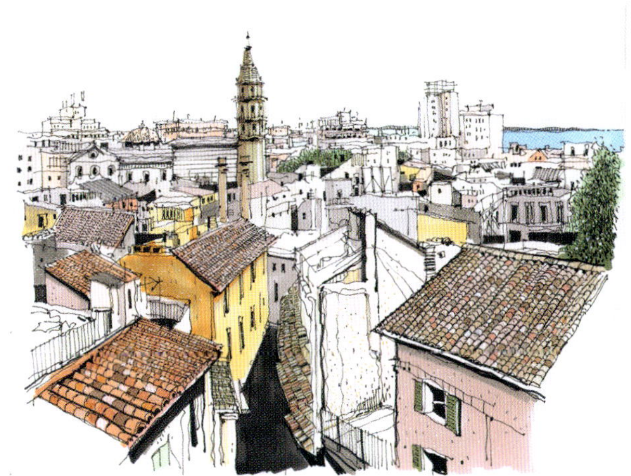

Palma, Majorca, Spain

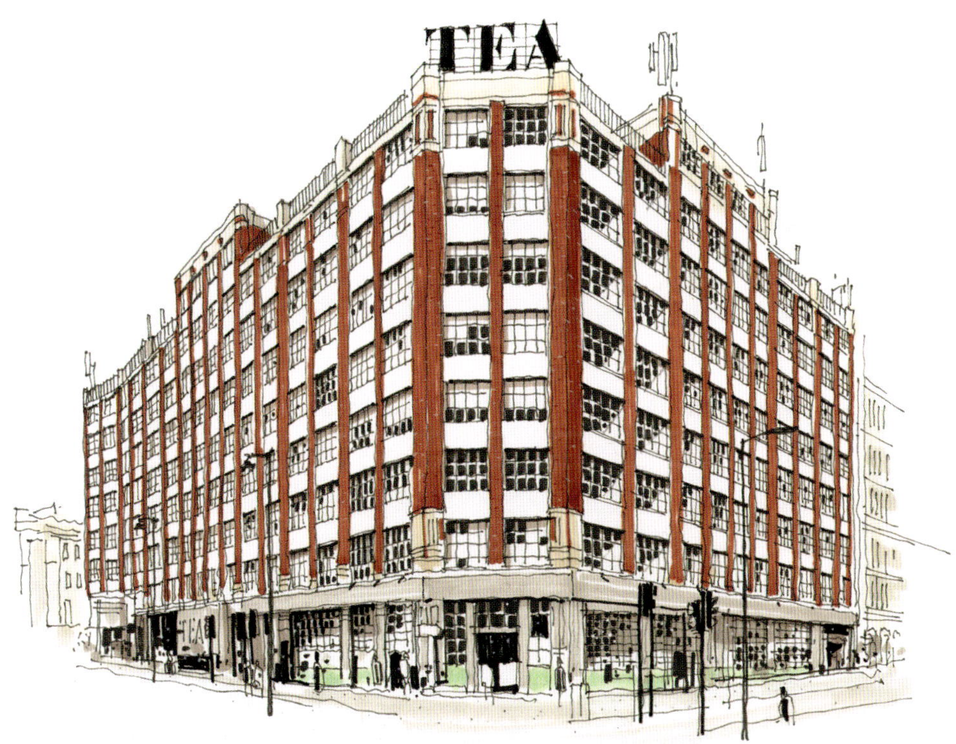

The Tea Building, formerly a Lipton Tea factory, Shoreditch, London, UK

PEOPLE IN PERSPECTIVE

When drawing in perspective, it's important to include people drawn at the correct scale.

The process is the same as drawing anything else in perspective: first, establish the eye-level and vanishing point. Place one standing figure with the head on the eye level line. Then draw receding lines from the top of the head and from the feet back to the vanishing point. These lines create a zone that determines the correct height of any other figures in the scene, no matter how far away they are.

As a guide, the average adult figure is around seven-and-a-half to eight "heads" tall. This helps when blocking in proportions, especially if you are less confident drawing people. Of course, not everyone is the same height, so this is just a rough guideline – but it's a helpful trick for showing perspective.

Even a simple outline or silhouette can provide an effective sense of scale within a scene. Remember also that figures closer to the viewer should overlap and show more detail, while those further away will appear smaller and less defined.

Personally, I don't include detailed figures in my own work – I've never felt especially confident drawing them. Instead, I usually use simplified forms to give a clear sense of scale. Adding small details such as bags, suitcases, or clothing can make even simple figures look more convincing. As with most things in drawing, practice is the only way to make figures appear more realistic!

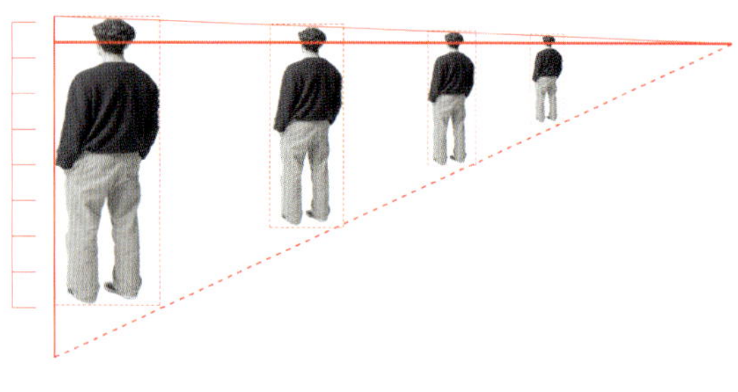

Diagram showing receding lines from the top of the head to the vanishing point for multiple figures. The scale adjacent represents approximately eight "heads" high.

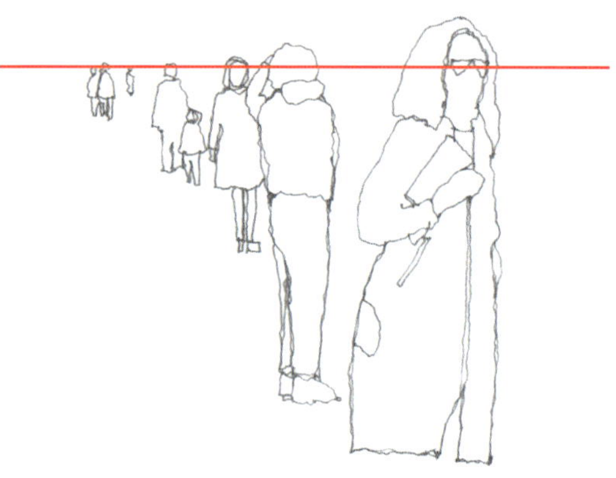

This is a good example of heads aligning roughly at eye level, regardless of the size of the figures or their position within the perspective view – excluding, of course, children and dogs. As with many perspective drawing principles, the rule is not absolute; sometimes it is simply a matter of trusting what your eye tells you.

FORESHORTENING

If you draw detailed figures, you need to consider foreshortening. When a body is viewed from above or below, it appears compressed. Imagine a vertical line showing a person's full height: from straight on (for example, see the figures on the opposite page), it shows the true height, but from an angle, it looks shorter. Understanding this principle will help your figures look more realistic from different viewpoints.

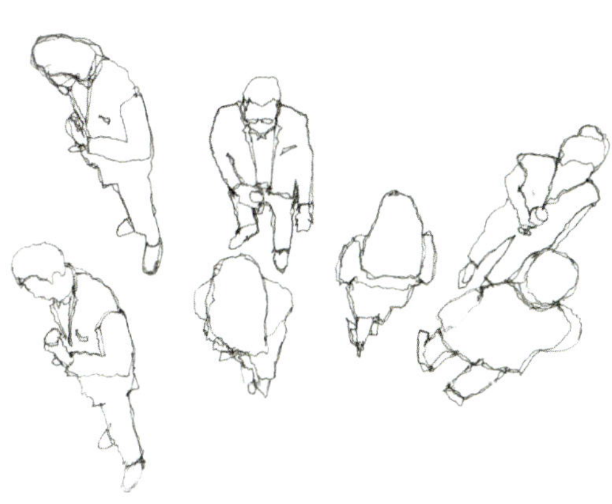

EYE LEVEL POSITION

Another important factor is the position of the eye level, which changes how figures appear. From a high eye level, looking down, people look more foreshortened and you may see mostly their heads and shoulders. From a low eye level, you see more of the undersides of forms, and horizontal features are emphasized. Always check that your figures are in proportion to their surroundings, such as doors, windows, and furniture.

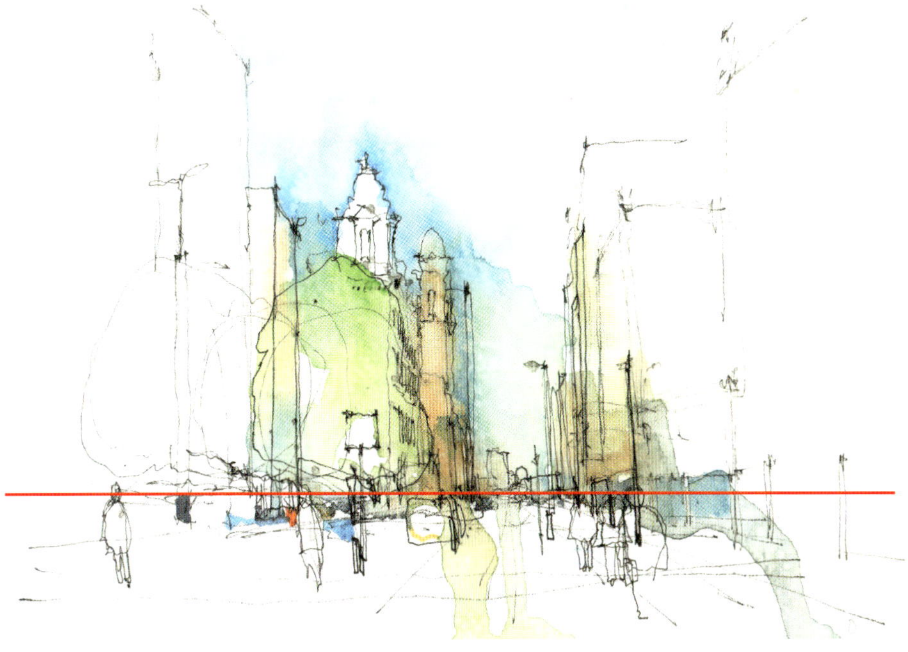

This loose sketch of a Manchester street is a good example of where I have drawn a few people for scale.

FEATURED ARTIST

Karen Jones

Manchester, UK

I am inspired to draw large-scale cityscapes and complex three-dimensional scenes, and particularly love the challenge of multi-point perspectives. I am also on a journey of self-discovery in how I express my response to my built environment in pen and watercolour. Being an architect this, to me, is finding the right balance between a technical drawing and a piece of art. The balance between structure and fluidity.

My technique has therefore developed over the years to draw fairly accurately in three dimensions in pen and then to soften the sketch with an emotional response in watercolour. Being automatic with the basics of perspective also helps me bend and manipulate it when I feel playful.

My preference for a successful sketch is in taking the time to study the scene initially, decide what I want to draw, its composition, and then to set it up in a light pencil. Being patient at the start pays dividends. Although it feels like a struggle in the beginning, the drawing work does becomes automatic.

I work with a pencil, a fountain pen, good-quality paper, and watercolour paint. I use a simple scaling technique using a pencil as the measuring tool and setting out a square grid (in pencil) on my paper. I position the horizon line and vanishing points and scale, roughly, the dominant shapes to my paper. I then draw in details, all relative to the vanishing points and horizon line, with smaller buildings and details drawn scaled relative to the larger shapes.

When I think the drawing work is complete, I assess how I feel about the experience and apply a limited colour palette in watercolour to try to capture that feeling.

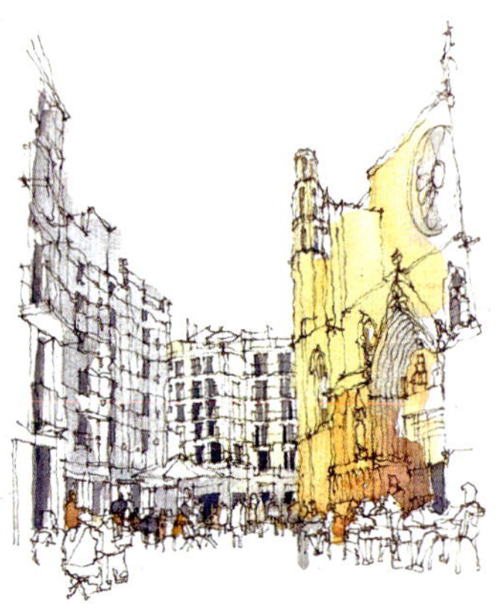

Placa de Santa Maria, Barcelona, Spain. From the tenth anniversary of the Barcelona Urban Sketchers Symposium celebrations meet-up.

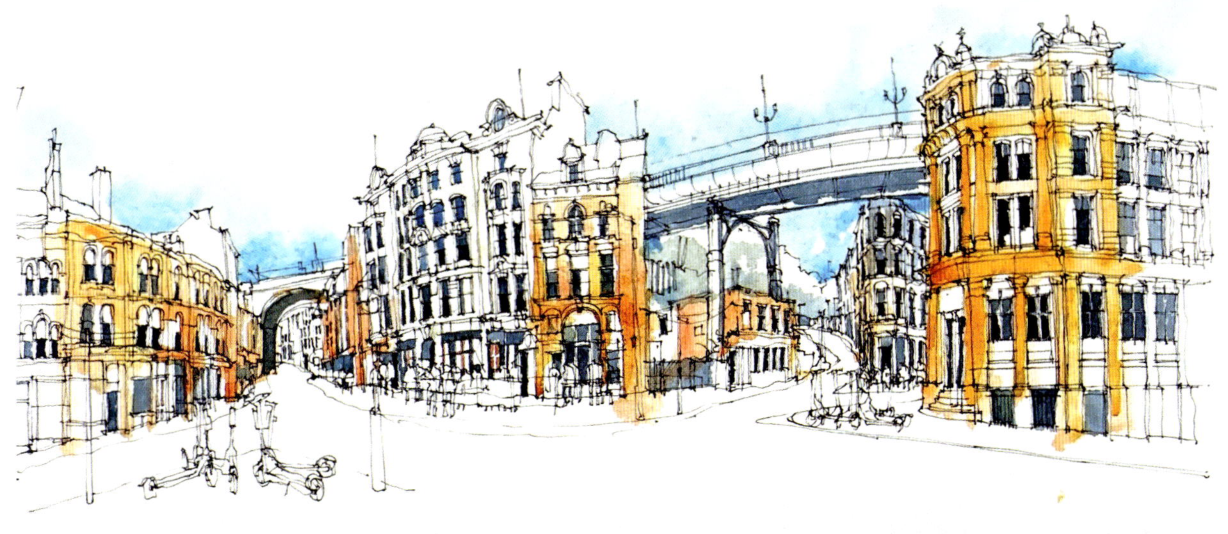

Sandhill/Queen's Street, Newcastle, UK

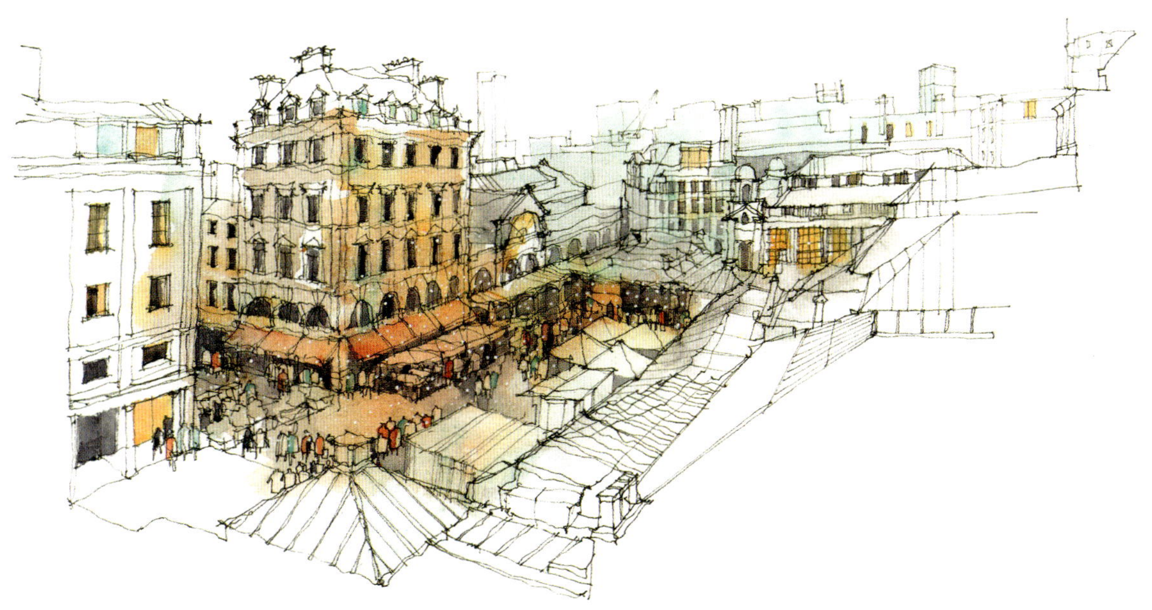

Covent Garden, London, UK

NO PERSPECTIVE

It's important to recognize that sometimes a scene offers little or no perspective viewpoint. This happens, for example, when we're looking directly at a building façade or the flat vertical surface of an object close to us. There are no obvious vanishing points or indications of foreshortening, or elements such as buildings receding into the distance.

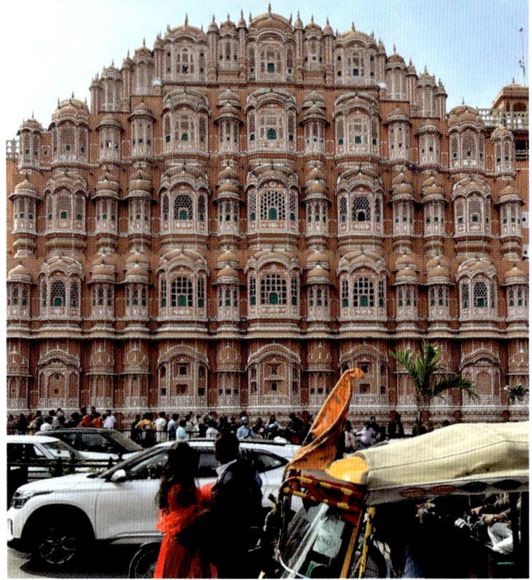

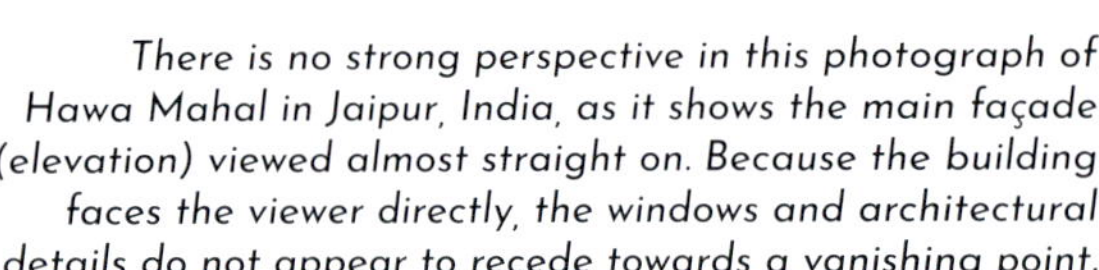

There is no strong perspective in this photograph of Hawa Mahal in Jaipur, India, as it shows the main façade (elevation) viewed almost straight on. Because the building faces the viewer directly, the windows and architectural details do not appear to recede towards a vanishing point.

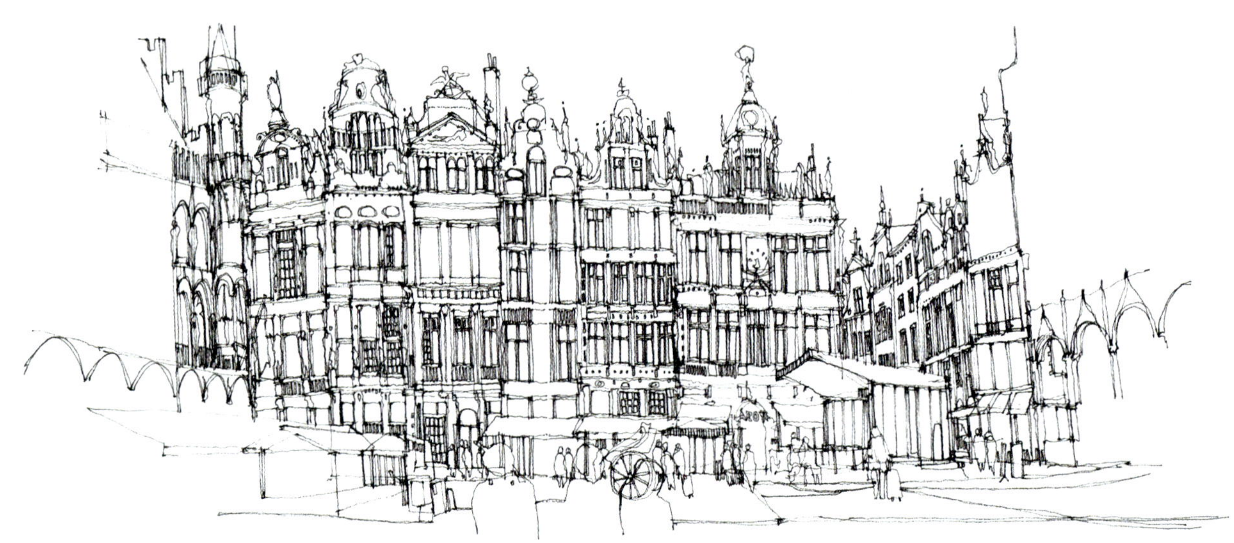

In this sketch of the Grand Place in Brussels, Belgium, you'll notice a subtle suggestion of perspective along the edges, but these are just compositional devices that guide the eye towards the central focus of the drawing.

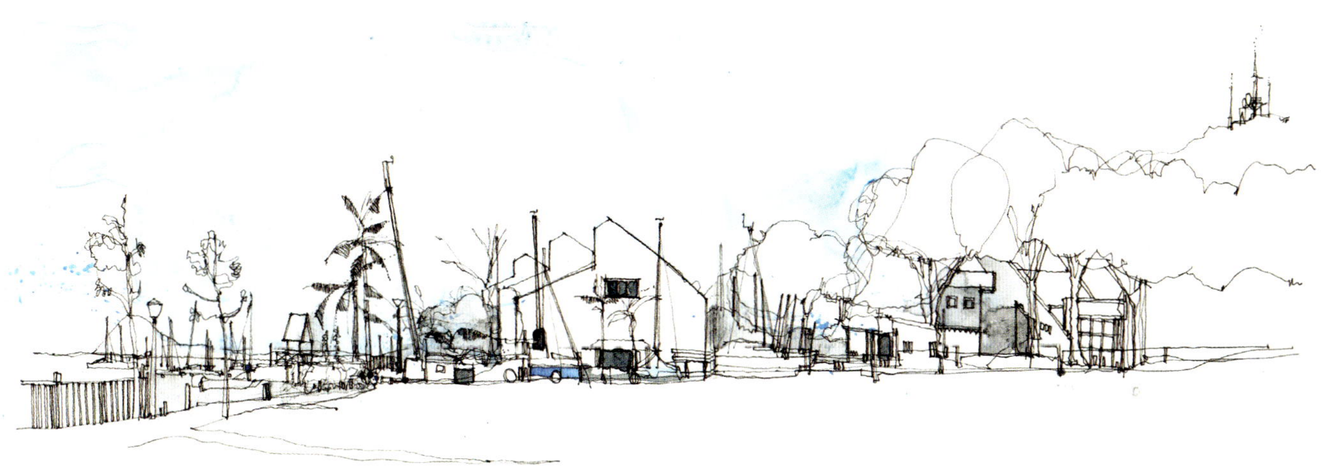

Here's another example of a sketch with limited or no perspective, showing the Changi Sailing Club, located at the far eastern tip of Singapore. The sketch itself is almost completely flat – more like an elevation – capturing a broad panorama from the Straits of Singapore on the left, across the sailing club apartments and dry-docked yachts, to the satellite masts of Singapore Air Traffic Control on the far right. The only real suggestion of depth or perspective comes from the use of heavier line-weights and a touch of colour in the foreground.

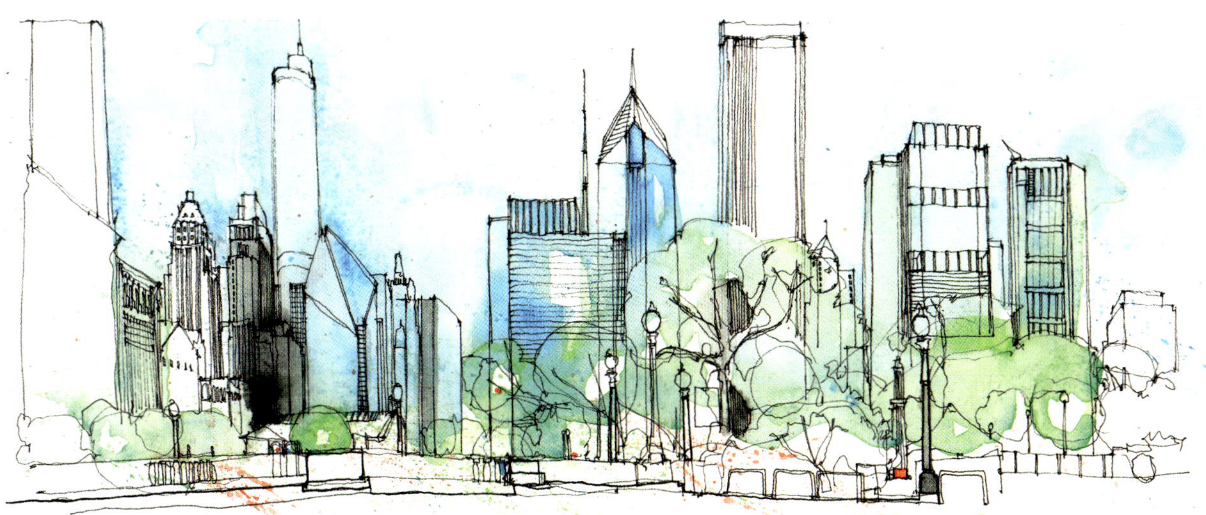

There's no clear perspective in this loose sketch of the Chicago skyline drawn from Buckingham Gardens. However, it does employ some of the same techniques demonstrated in The City is a Stage section. The drawing is constructed in layers – the park in the foreground, the skyscrapers immediately behind it, and a more distant tier of high-rise buildings – all unified by a tonal watercolour wash that overlays each layer.

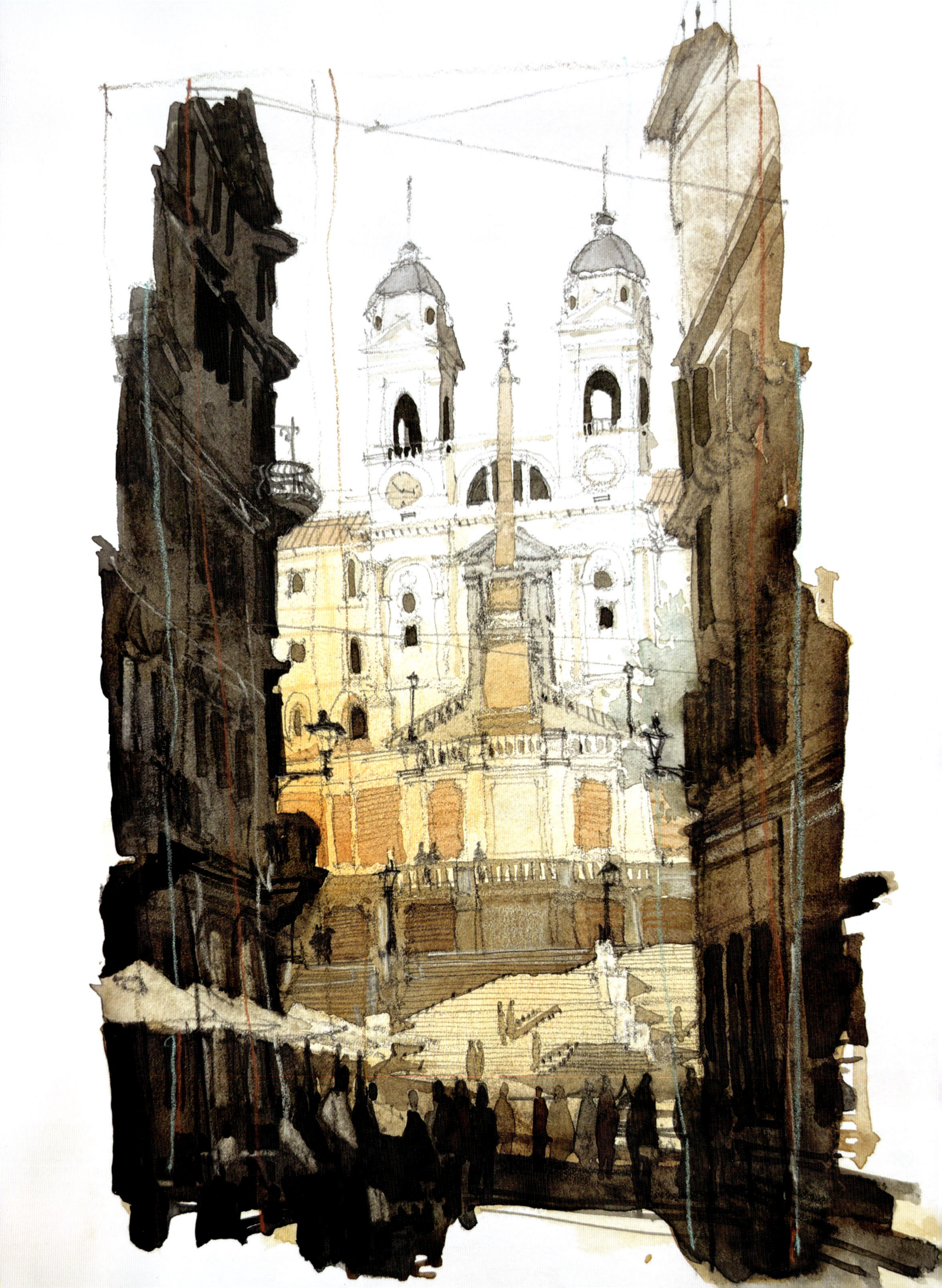

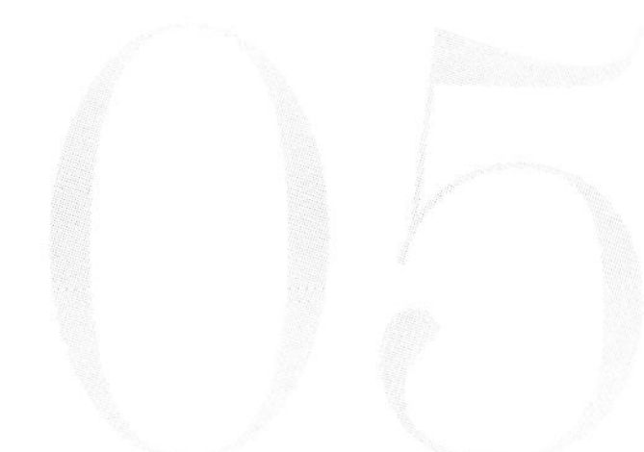

DIAGONALS, ARCHES, AND ELLIPSES,

This chapter covers one of my favourite subjects: how what I refer to as diagonals – they are effectively diagonal lines – can help us work out perspective. I always tell my students it feels a little bit like magic, even though it's firmly grounded in mathematical principles.

By applying diagonals, we should be able to solve even the most complex perspective problems. They are particularly valuable when it comes to drawing arches, circles, and ellipses – shapes that are notoriously difficult to construct in perspective.

The use of diagonals give us a clear, reliable method to approach these challenges. They also provide a simple way to find midpoints within objects that recede towards a vanishing point.

In this chapter, I explain the basic principles, and I return to their use later in Chapter 06, where diagonals are applied to one- and two-point perspective grids.

A one-point perspective view of the Piazza di Spagna in Rome by Indonesian architect and illustrator Darman Angir

USING DIAGONALS

In Chapter 02 we touched on how to achieve the correct depth in a one-point perspective view. Now, let's introduce a more effective method using this simple technique that helps establish the right proportions in perspective.

Here's a one-point perspective drawing, with the shaded section highlighting the shopfront (A). Can you pinpoint its midpoint? Determining this by eye alone is quite challenging but give it a try.

Now, to find the centre point more accurately,' first treat the building's frontage as a trapezoidal plane (B). Then, draw diagonals from each corner and where they intersect marks the midpoint. From this point, draw a vertical line to divide the trapezoid into two equal halves in perspective (C).

Interestingly, as you can probably see, this midpoint is probably different to what your eye sees (or perceives). This is because foreshortening in perspective alters proportions, reminding us that accurate construction sometimes conflicts with visual intuition.

Note

In architecture, trapezoidal describes a shape similar to a trapezoid – a four-sided figure with at least one pair of parallel sides. It usually refers to the form or geometry of a building element or space.

A

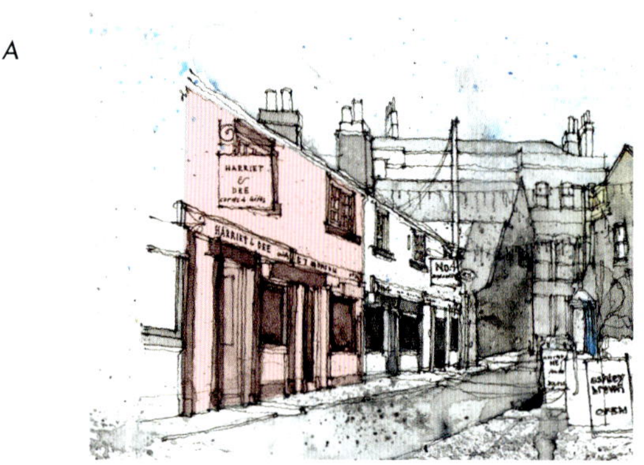

B

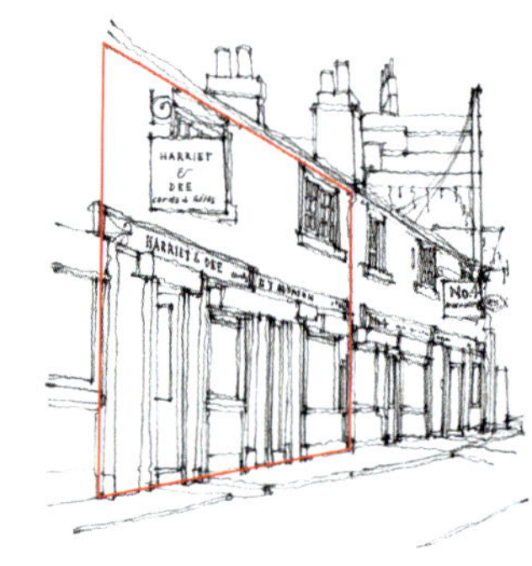

C

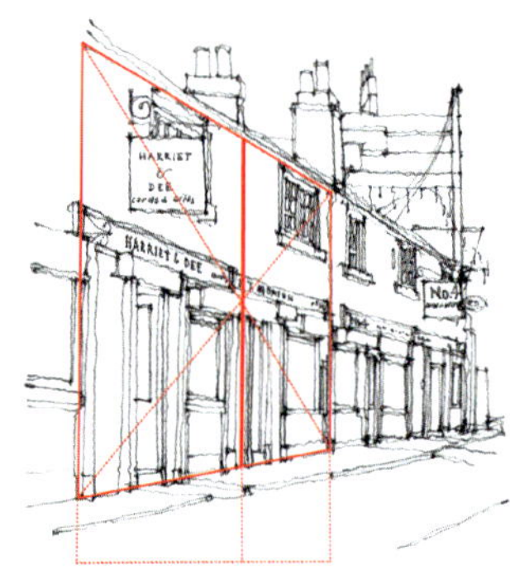

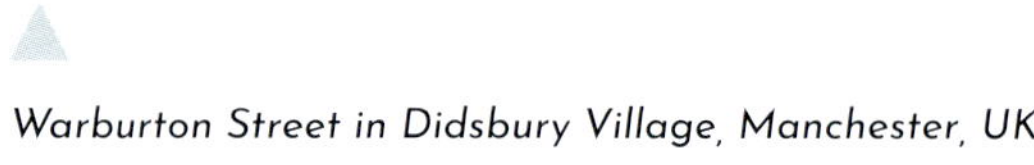

Warburton Street in Didsbury Village, Manchester, UK

Using diagonals is a simple but effective method for finding the centre of a building in perspective, which I often use when placing a centrally located entrance or another key architectural feature. In my sketches, I tend to apply it quite loosely – you can sometimes see the faint diagonal guidelines where I'm working out the midpoint of a building as it recedes through foreshortening.

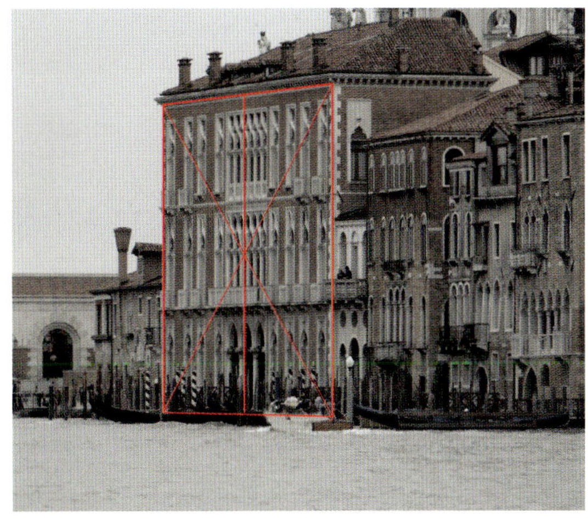

This photograph of a palazzo on the Grand Canal, Venice, Italy, illustrates how theory translates into practice. By applying diagonal lines, I located the vertical midpoint of the façade, which aligns precisely with the central pointed arch of the first-floor loggia, and with the centre of the three-arched loggia at water level.

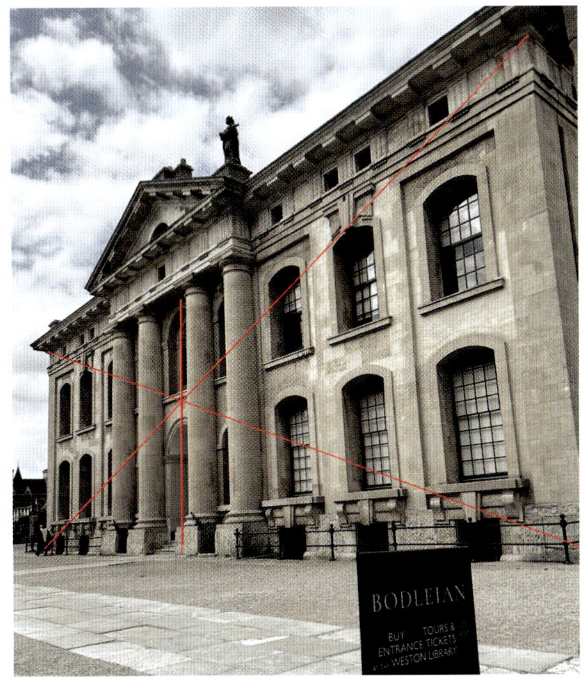

This photograph of the classical façade of the Bodleian Library in Oxford, UK, also illustrates this principle beautifully. The two diagonal lines pinpoint the exact centre of the building, confirmed by the arched opening above the main entrance.

METHOD 1

Let's break this down into a series of diagrams to show how diagonals work in perspective drawing and how they help divide space. Here I illustrate this both on a flat plane (A) as well as in a perspective viewpoint (B).

1. Draw a rectangle; then using diagonals, work out the midpoint. You have divided the rectangle in half.

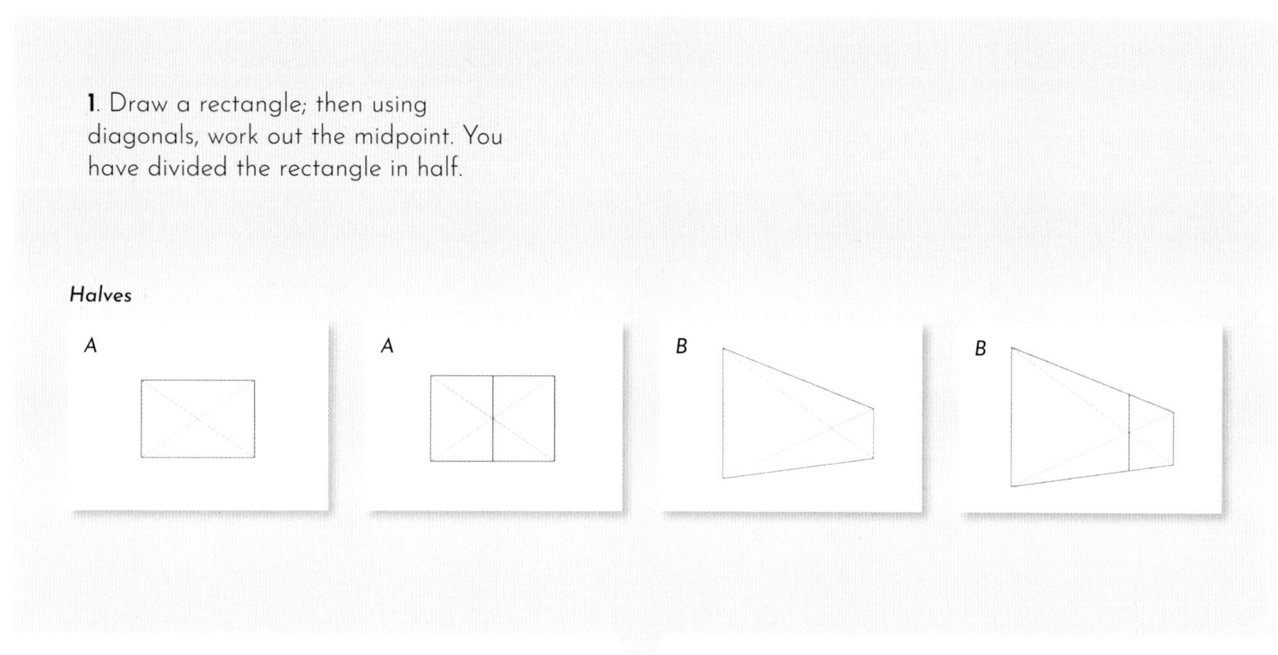

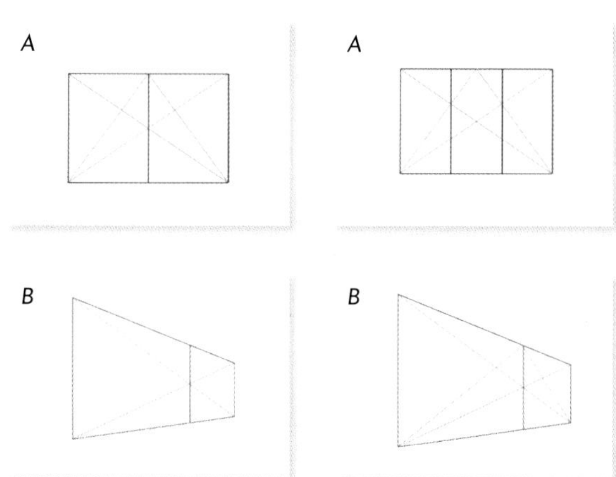

2. Draw additional diagonals as shown here, from the top midpoint to the bottom corners – using these intersection points gives you thirds.

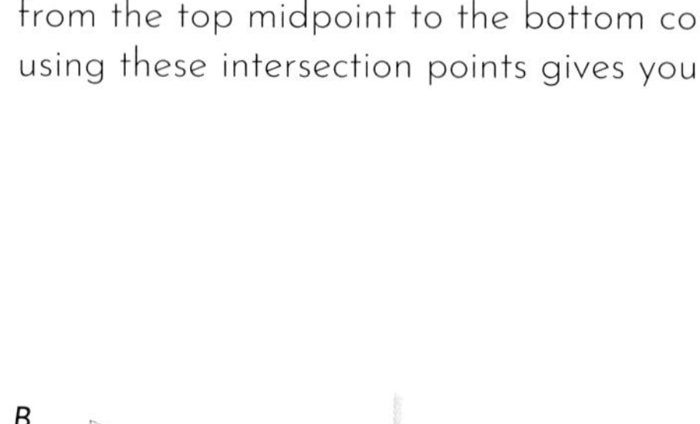

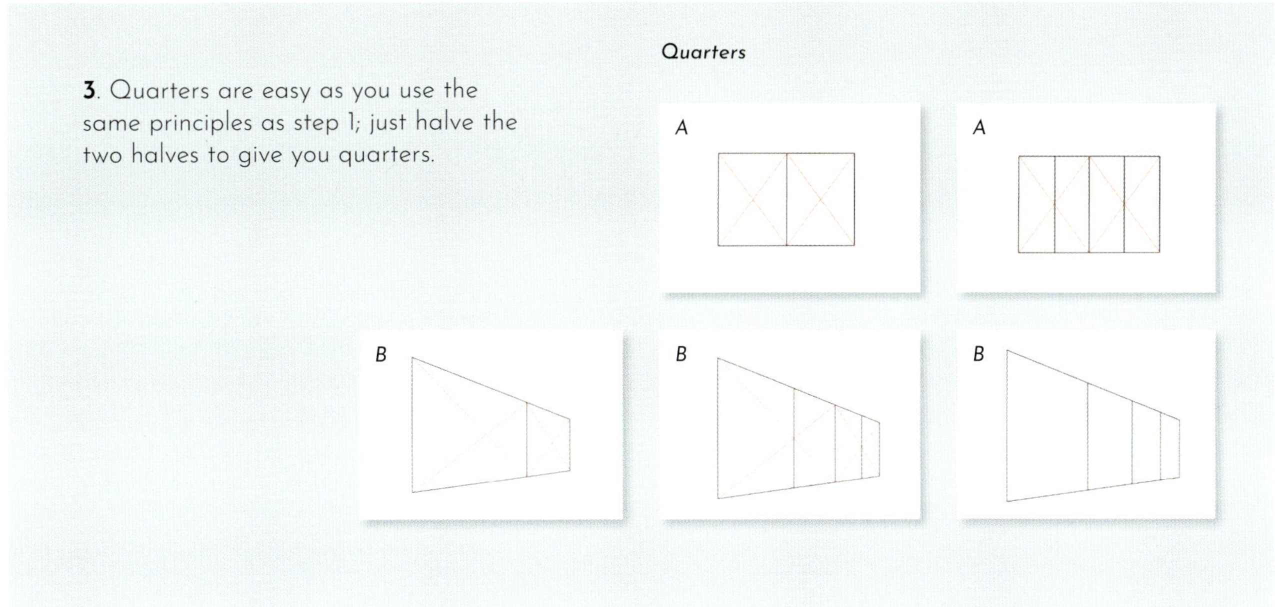

3. Quarters are easy as you use the same principles as step 1; just halve the two halves to give you quarters.

We can continue subdividing our trapezoidal shape indefinitely using this technique. It's a great addition to your perspective drawing toolkit – and it is important to note that the same principles apply if we subdivide horizontally instead of, or as well as, vertically.

This photograph of St Thomas's Basilica in Chennai, India, is taken from a low viewpoint. It illustrates how a façade can be divided into thirds, where the divisions align clearly with the columns. Although it is not strictly a perspective view, it effectively demonstrates the theory discussed here.

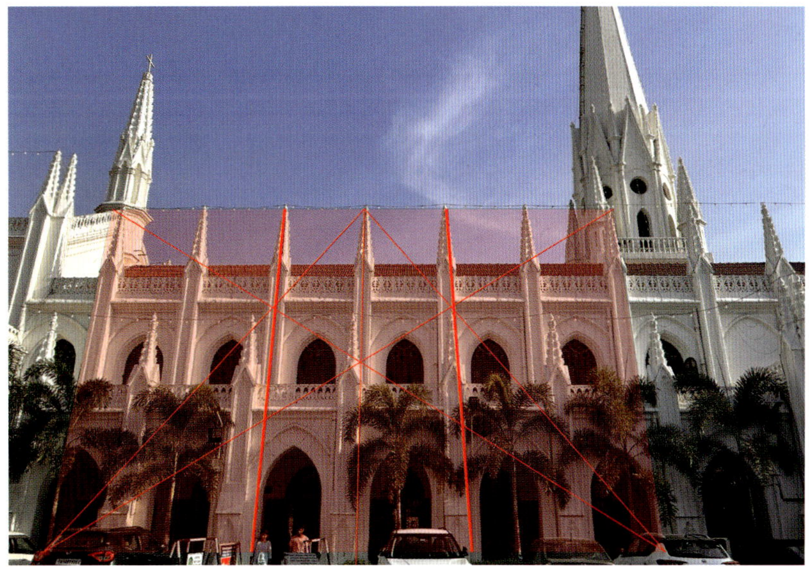

METHOD 2

Here's another method for using diagonals in perspective drawing. This example uses a vertical plane, but the same principles apply horizontally.

In this case, we will divide a trapezoidal shape into six equal vertical sections.

1. Draw your trapezoidal shape and eye level, then mark the vanishing point (VP). Identify the vertical edge that is furthest away from the vanishing point. Measure it carefully and divide it into six equal parts.

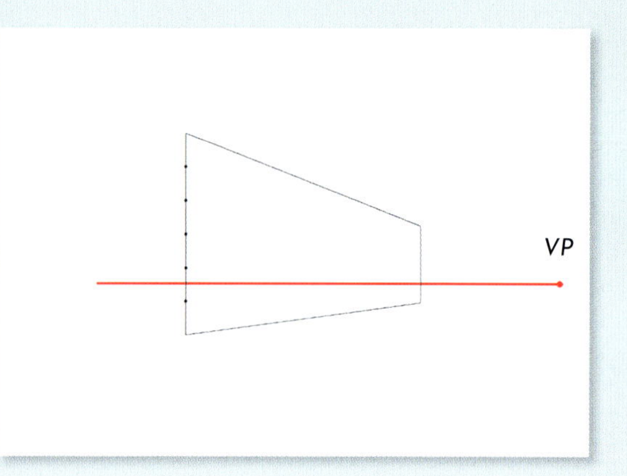

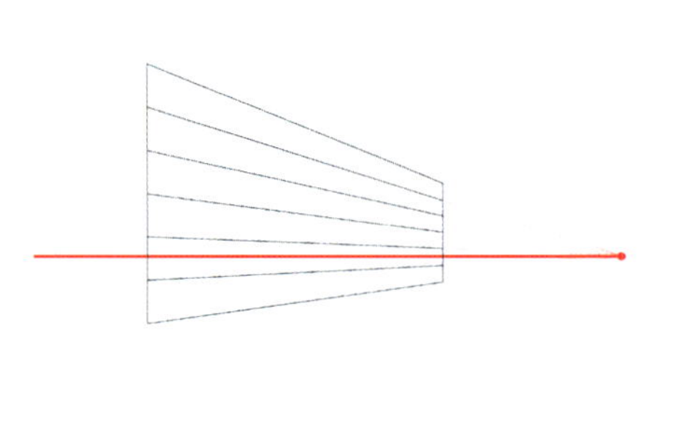

2. From this edge, draw lines that extend to the vanishing point from each of the marked divisions.

3. Now for the clever bit! Draw a diagonal line from the top right corner to the bottom left corner and mark the points where they intersect with the receding lines back to the vanishing point.

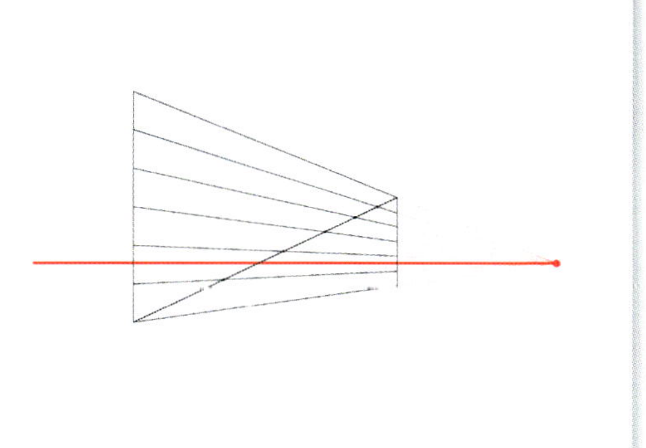

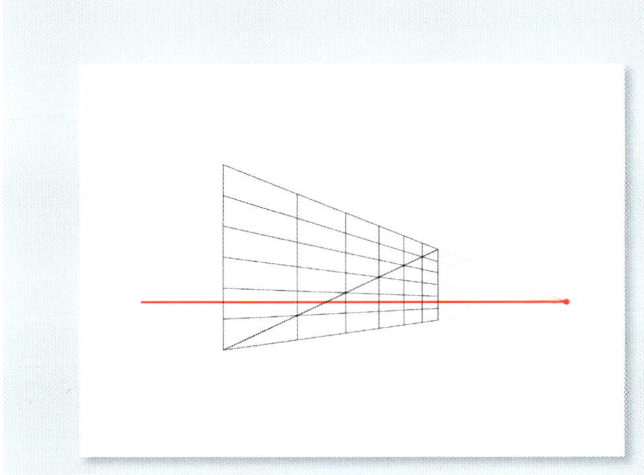

4. Wherever these intersections are, draw vertical lines. If your measurements are accurate, these verticals will divide the trapezoid into six equal sections – perfectly in perspective.

Tip: As a check, over this final diagram, I have added diagonal lines from the corner points of the original trapezoidal shape and you can see they have intersected on what I have worked out as the midpoint. So it's good to know you can use different methods to get to the same result!

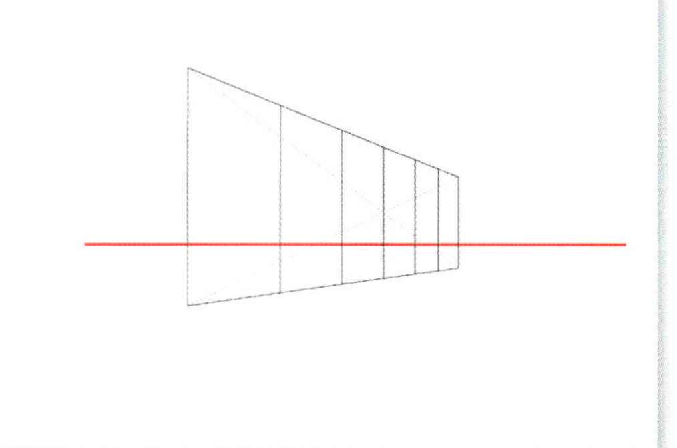

ARCHES

Arches can be one of the most challenging elements to draw accurately in perspective, but applying the principles discussed earlier – such as using diagonals to find the centre point of trapezoidal shapes – will help you draw them accurately in one-point perspective.

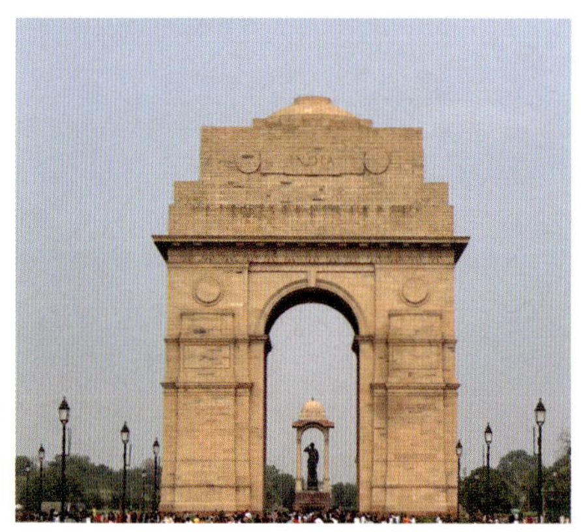

This photograph shows the iconic India Gate in New Delhi. Captured in clear one-point perspective, the arch perfectly frames Netaji's Statue beyond, with the springing line (the starting point of the curve on an arch) of the arch clearly visible.

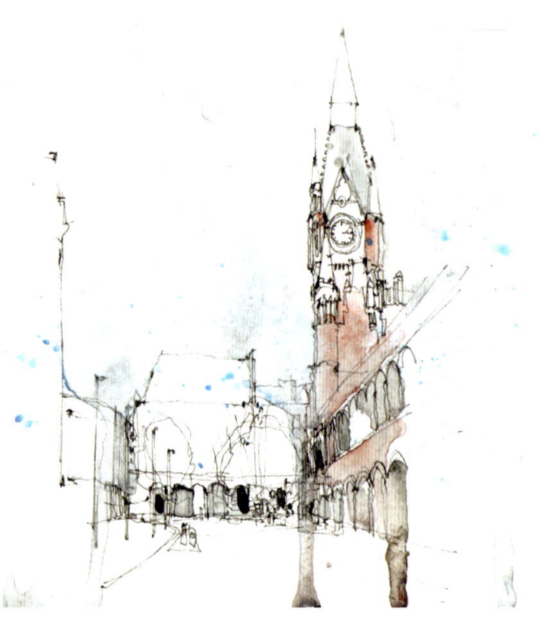

This quick sketch on location of the arches at St Pancras International, London, suggests perspective and depth in an unstructured way yet still conveys the sense of recession.

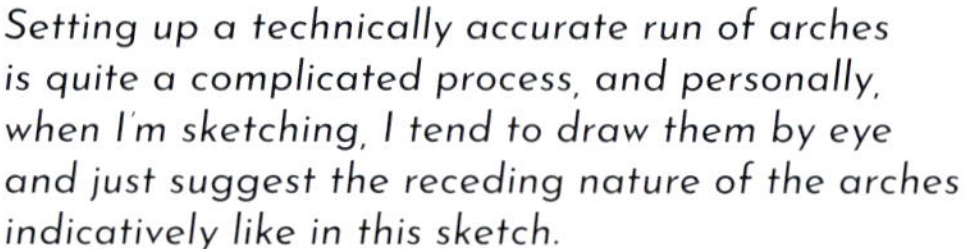

Setting up a technically accurate run of arches is quite a complicated process, and personally, when I'm sketching, I tend to draw them by eye and just suggest the receding nature of the arches indicatively like in this sketch.

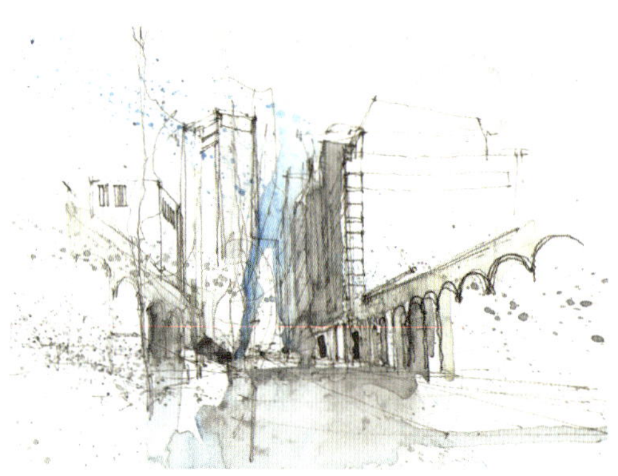

IN PRACTICE

I've used a photograph of Marble Arch, London, to show two different examples, each worked directly over the original photograph to demonstrate how the theory functions in practice. In example A, you can see how the larger central arch is constructed. First, I draw a rectangle in perspective that fits over the opening of the arch, with the top and bottom edges clearly receding towards the vanishing point (the top edge aligns with the springing point of the arch). Then, by adding diagonals, I locate the centre of the arch. When I project this point upward to meet the curve, it aligns perfectly – showing how the theory works practically.

I then repeat the same process for the two smaller, equally sized arches (example B).

Note

This principle also works for pointed arches (such as at London's St Pancras International) as well as rounded arches

A

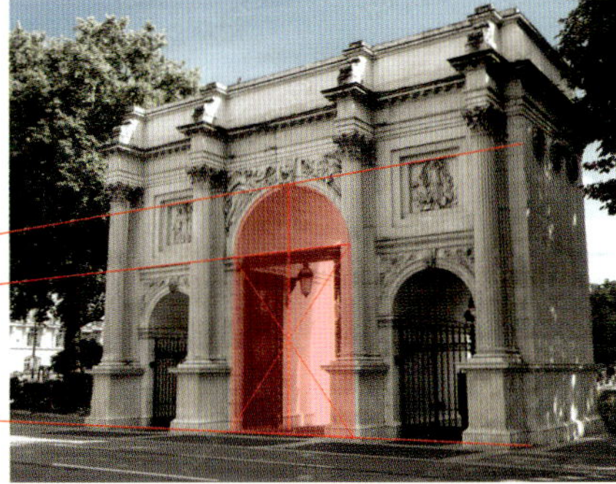

B

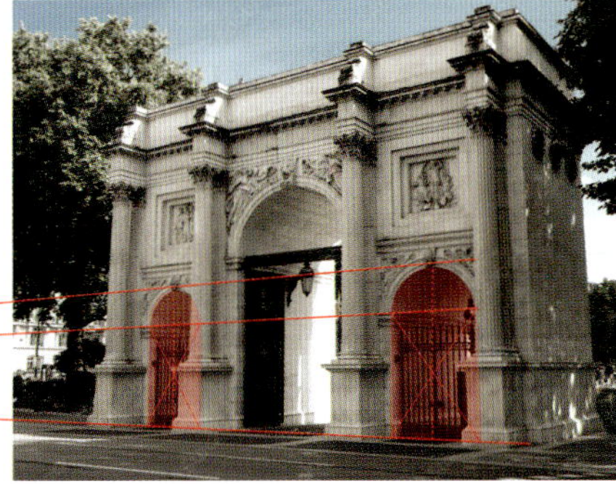

MULTIPLE ARCHES

This diagram, created in Adobe Illustrator, shows imaginary arches drawn in one-point perspective. The series of arches recede to the vanishing point. By projecting guidelines from the outer corners of the front central arch, we can establish the second arch as a rectangular shape positioned behind it. Using the arch nearest to us, we can extend the corner points back towards the second arch to determine its height. The second arch slightly overlaps the third arch as they recede to the vanishing point, but the principles are still the same.

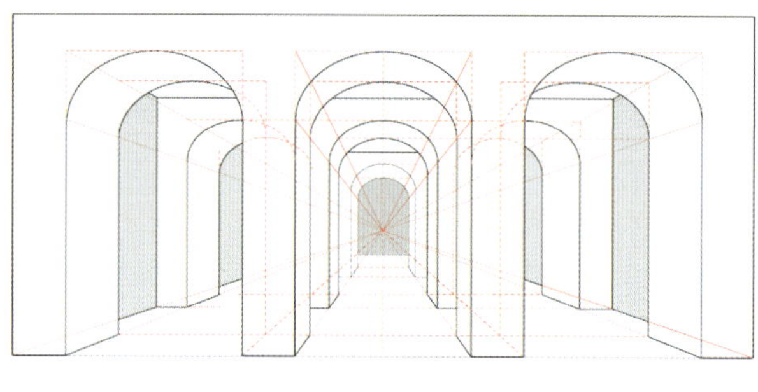

Diagram showing arches in one-point perspective

DRAWING A MORE COMPLEX FAÇADE

Now I'm going to demonstrate how the principles of diagonals can be used to draw a more complex façade with multiple arches. The example here is Poznań City Hall in Poland. For the purposes of this demonstration – and to keep things as simple as possible – I'll prioritize the open arches, rather than the solid arched side panels that you can see on this photograph of the building.

Note

I would never usually work out an arched façade in such detail, but I thought it was important to share the principles with you.

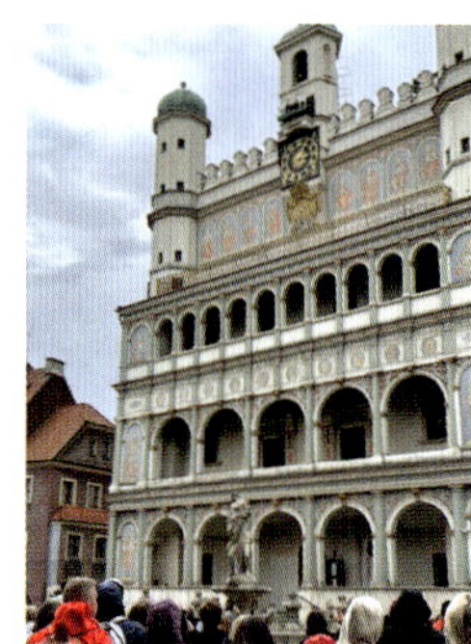

1. First, draw the trapezoidal outline shown in blue and take care to make sure the proportions look right. Include the left-hand vanishing point. While there is a right-hand vanishing point, as it is technically a two-point perspective view, nothing in this particular view appears to be receding to it.

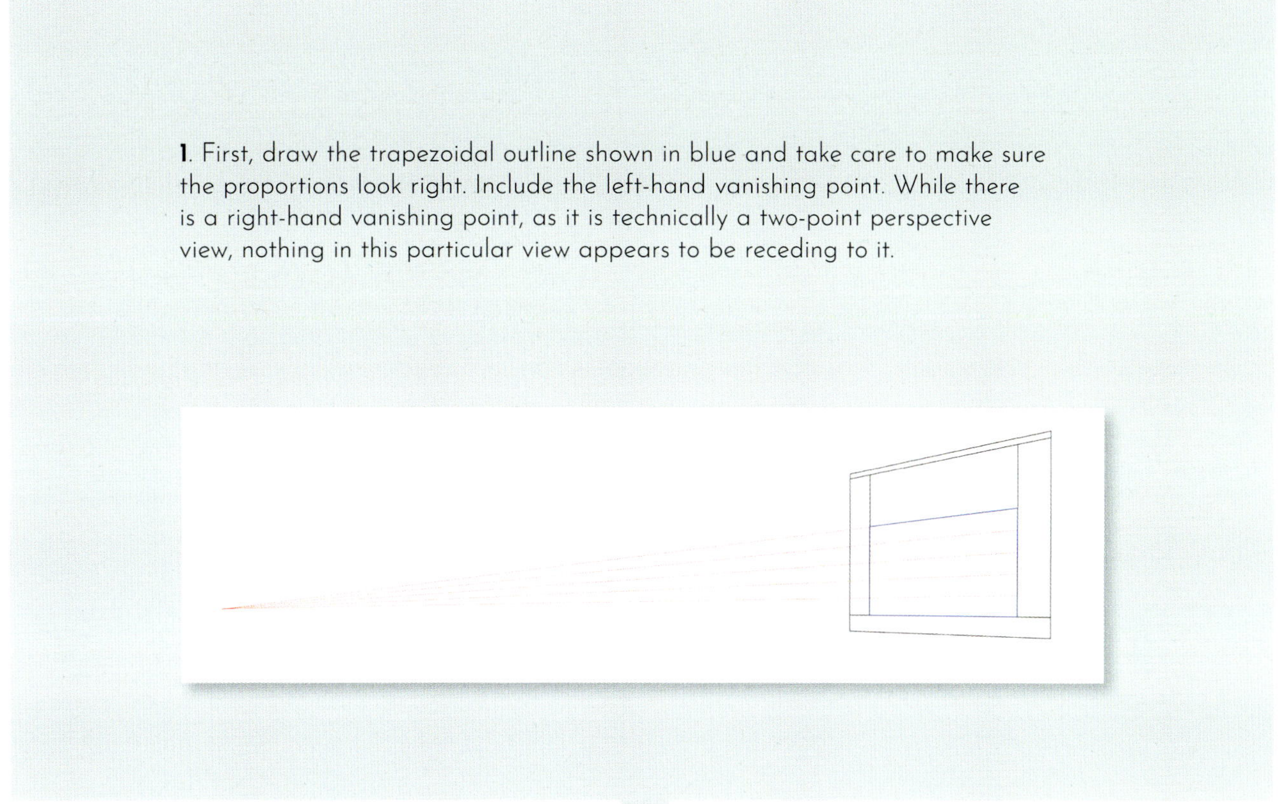

2. At ground and first-floor level, the façade is five arches wide, so divide the trapezoid shape into five equal vertical sections using a diagonal measuring method (see Method 2 in the Using Diagonals section). Starting from the right-hand front edge (the nearest vertical corner), mark each division and draw faint construction lines receding towards the vanishing point. These are guidelines only and are not part of the final drawing. Because the perspective recession is very shallow, the left-hand vanishing point lies to the far left of the page.

Now draw a diagonal line from the top left to the bottom right. Where this diagonal intersects the receding guidelines, draw vertical lines. This divides the façade into five equal sections, allowing you to map in the main architectural features in simplified form.

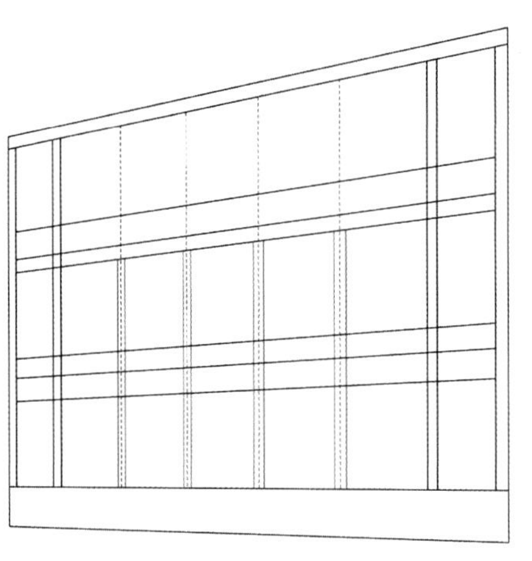

3. The four interior vertical lines created by this diagonal measuring method mark the centre of each column. Using these centre lines, draw the columns and give them depth, which in this example is estimated by eye. At this stage, you can also mark the zones of horizontal decorative banding, which represent the floor levels behind the façade.

4. Next, locate the springing lines for the arches at both ground and first-floor level, drawing them as receding lines back towards the vanishing point. This gives you a series of rectangles between the columns. Now the fun part: in each rectangle, use diagonal lines to find the centre of the rectangle then draw a vertical line to where it intersects with the horizontal decorative banding above; this gives you the centre of each arch – and you can now draw the arches.

Tip: When sketching arches, don't worry about perfect geometry – suggest the rhythm and flow instead. A few confident lines often capture the character better than overworking the details.

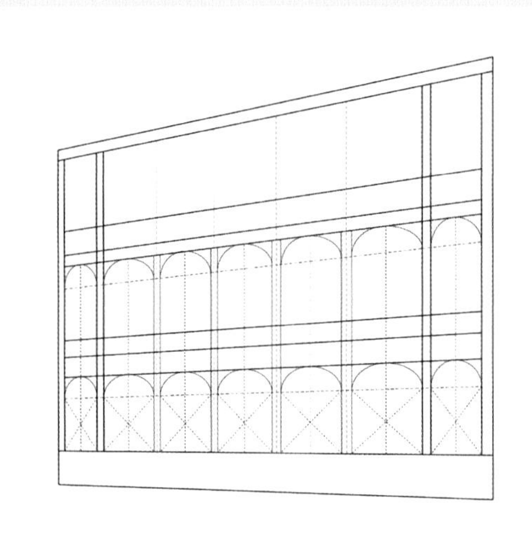

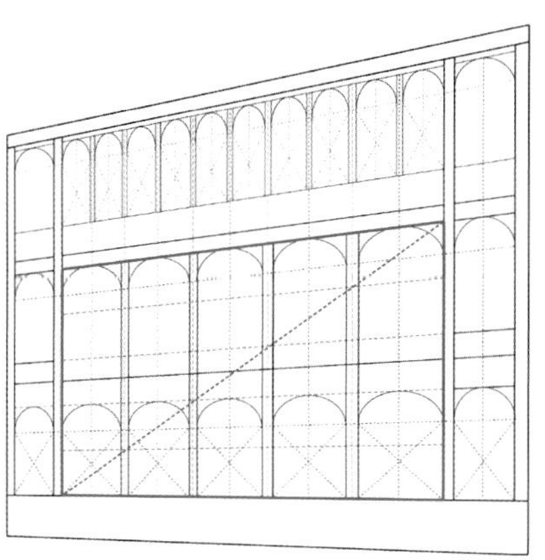

5. Now move to the second floor, where we have double the number of arches, and use diagonals again to set up these additional intermediate columns.

6. In this final diagram, I have highlighted the key geometric principles used to space the arches. The trapezoid shape is identified and divided into five equal columns at ground and first-floor level using a diagonal line. This establishes the position of the ten arches at second-floor level in correct perspective. You could also add the narrower solid arches at each side to complete the drawing.

Note

Although this is technically a two-point perspective view, we haven't needed to use the right-hand vanishing point to establish any of the geometry. However, it's still present, and in the original sketch, sketched on location, there is just some light indication in the depth of the arches.

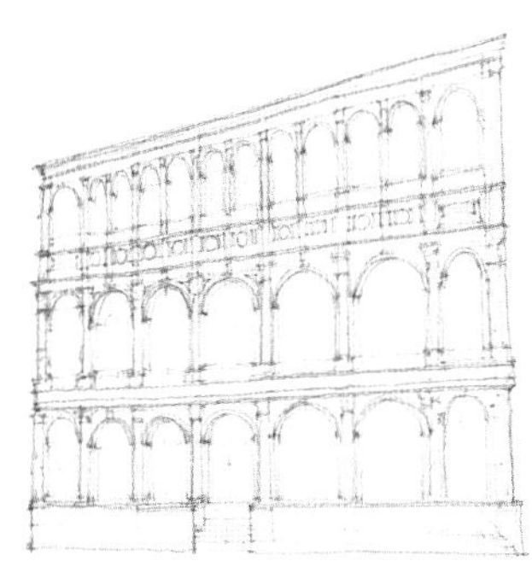

ELLIPSES

One of the most challenging parts of perspective drawing is getting curves and ellipses right. This becomes especially difficult in architectural drawings, where curved buildings and arches must be drawn accurately.

When we say ellipses, we mean circular shapes seen in perspective, so they appear elliptical, such as this round bowl that has been photographed in four positions. As the view of the bowl changes, the circular bowl appears to be elliptical. The closer the ellipse is to the eye level, the flatter it will appear.

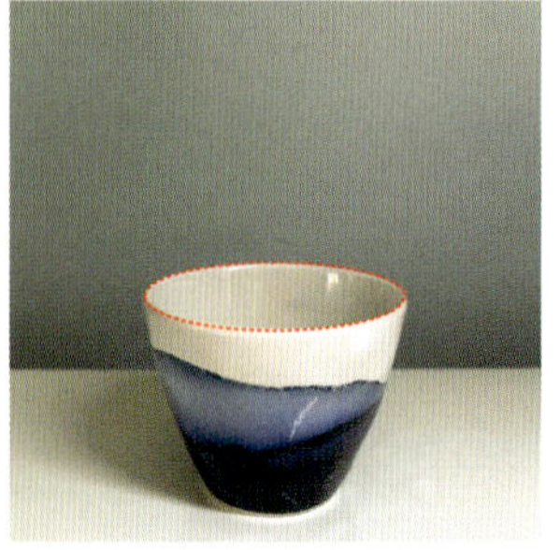
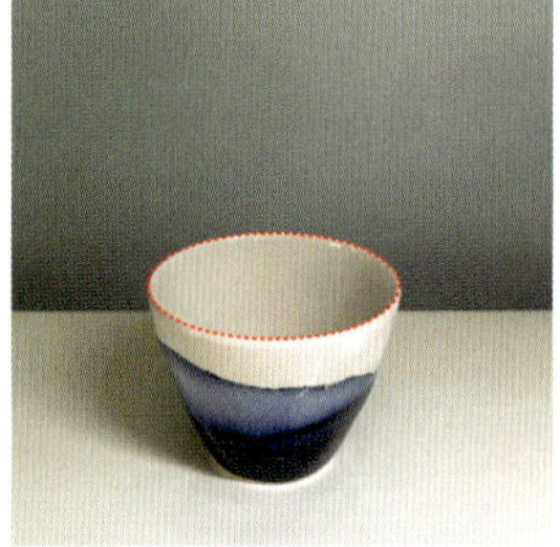
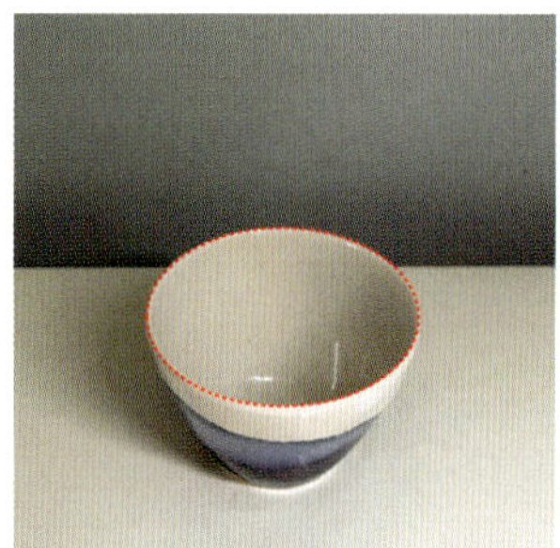
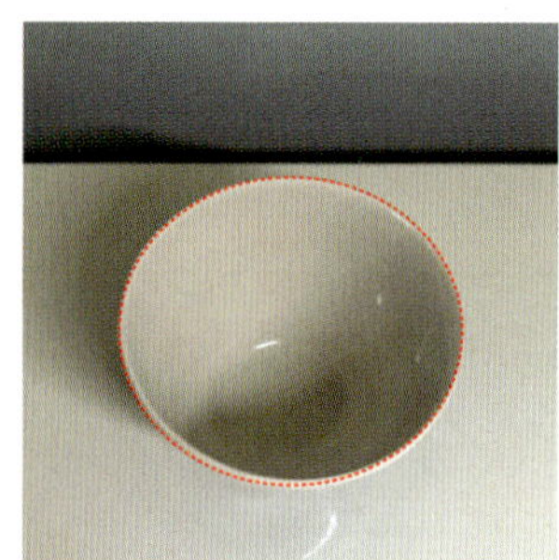

Note

At eye level, ellipses appear flattest. As you move above or below eye level, the ellipses expand and become more like circles.

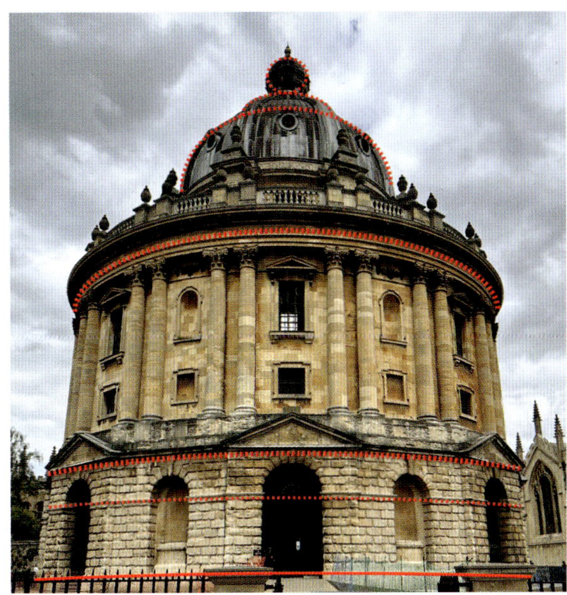

Taking this back into architecture, consider this photograph of the Radcliffe Camera in Oxford, UK: the building is designed in a series of horizontal layers. At eye level the ellipses appear shallowest, but as you move up or down from eye level, they open out. So, near the ground the curved elements are practically flat, while higher up the circular form of the building becomes increasingly clear against the sky.

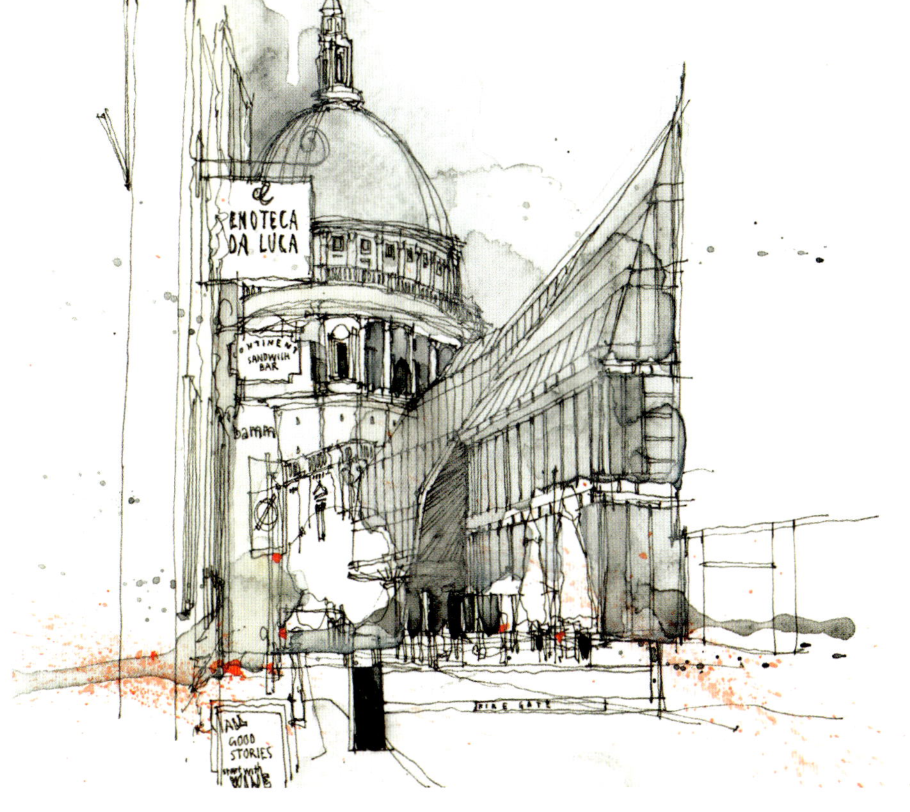

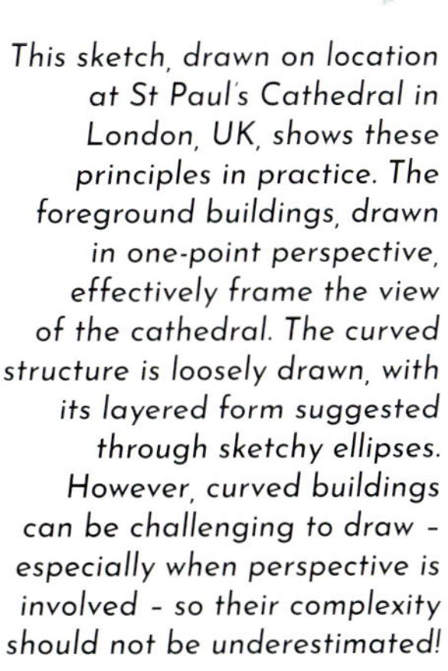

This sketch, drawn on location at St Paul's Cathedral in London, UK, shows these principles in practice. The foreground buildings, drawn in one-point perspective, effectively frame the view of the cathedral. The curved structure is loosely drawn, with its layered form suggested through sketchy ellipses. However, curved buildings can be challenging to draw – especially when perspective is involved – so their complexity should not be underestimated!

HOW TO DRAW A CIRCLE IN PERSPECTIVE

But back to the theory – let's look at drawing circles and ellipses in perspective. The principles are like those used for drawing arches, and rely on diagonal lines. In this example, we'll be focusing on *horizontal* circles and ellipses, rather than the *vertical* ones used for arches.

First, remember that a circle drawn in perspective will appear as an ellipse. To construct it accurately, we use what we call the "framework technique" to create a "crate" or box around the circle.

1. Here, you can see a horizontal trapezoid shape – this represents a square (or box) receding to two vanishing points (VP1 and VP2). By using diagonals, we can locate the centre of the circle. From there, lines drawn from both vanishing points through the centre and extended to the outer face of the trapezoid establish four key points on the circle's perimeter.

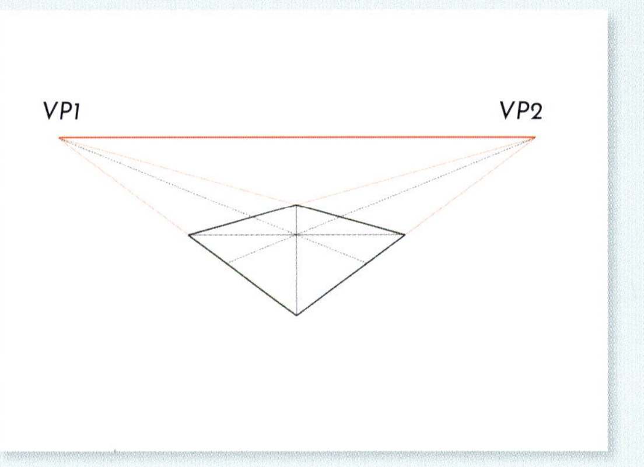

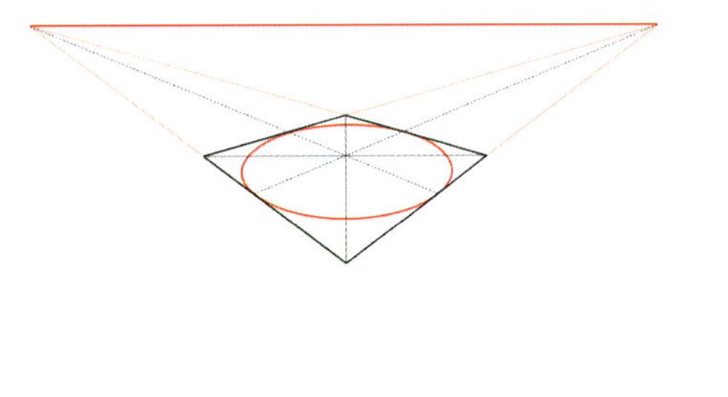

2. Using these points as guides, the ellipse can then be sketched freehand (although this diagram has been created digitally).

3. Next, we can extend the framing box upward to extrude the circle into three dimensions – a useful method for drawing cylindrical or rounded forms, such as furniture, in perspective. As the circle is extruded, notice how the plane of the ellipse changes. This mirrors what we observed in the bowl photographs shown previously. Remember, the closer an ellipse is to eye level, the shallower it appears.

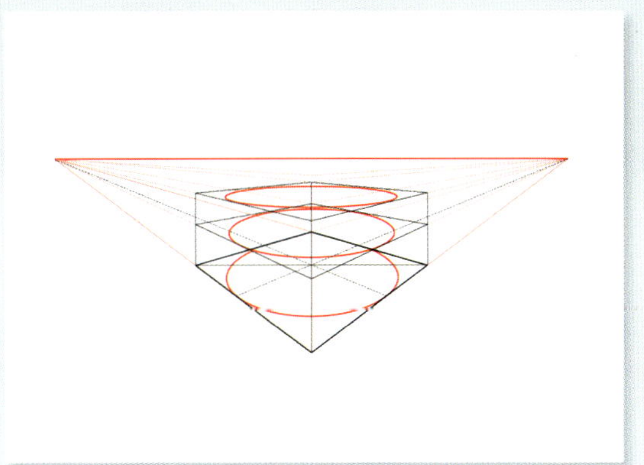

In these two diagrams, I've shown how circles are drawn in both one-point and two-point perspective. In the one-point example, the ellipses are shown in plan and elevation – horizontally and vertically. I start by using a framing rectangle to establish the boundaries of the circle. Then, by drawing diagonals, I can identify the four key points (north, south, east, and west) where the circle in perspective touches the frame. For clarity, I created these diagrams in Adobe Illustrator, but the same method can be hand drawn or using any digital drawing program.

In the two-point perspective diagram, the same principles apply, except that the elevations recede towards separate vanishing points. The construction method, however, remains essentially the same.

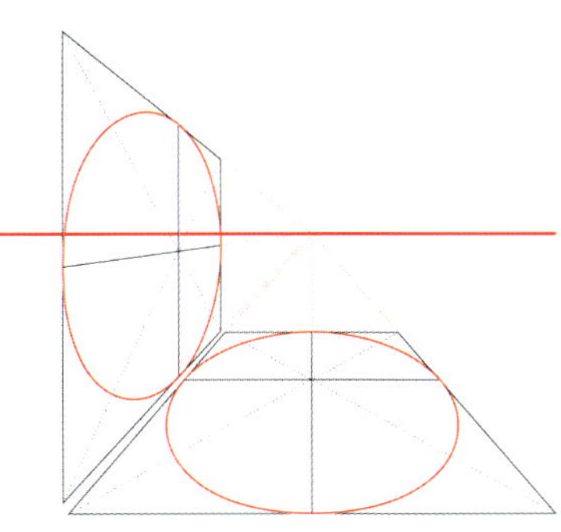

Diagram showing horizontal and vertical circles (or ellipses) in one-point perspective

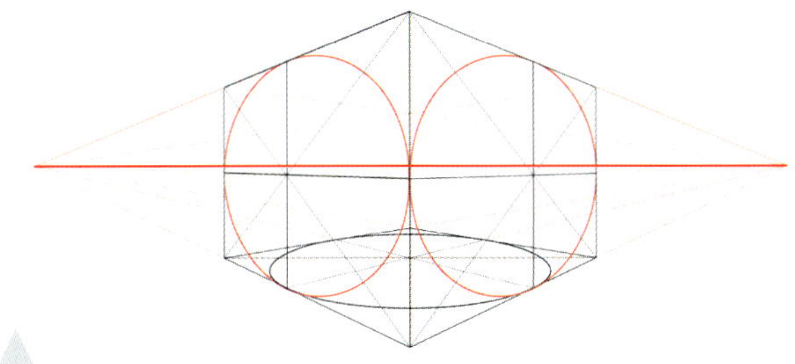

Diagram showing horizontal and vertical circles (or ellipses) in two-point perspective

FEATURED ARTIST

Stephanie Bower

Seattle, USA

I sketch to learn about and remember the architecture I see. That means, yes, I count windows and floors, as some degree of accuracy is important as I attempt to understand the architect's intent.

To accomplish this, instead of starting with what catches my eye the way that most artists start, I think like an architect and start by drawing big shapes. I ignore the lovely details and reduce what I see to simple squares and rectangles, and then locate my vanishing point and eye level relative to those big shapes, all drawn in barely visible pencil lines on my paper. Next, I build the perspective view in layers by breaking down those big shapes into smaller shapes and finally add the details last. I love these pencil drawings, so when it's time to add watercolour, my goal is to achieve a happy balance between line and colour in which both work together.

The best part is that years later, I'll look at a sketch and remember everything – the building, the location, the sounds I heard while working, even the feel of the breeze and the sun. Sketching is so powerful, it captures so much more than the click of a camera ever could.

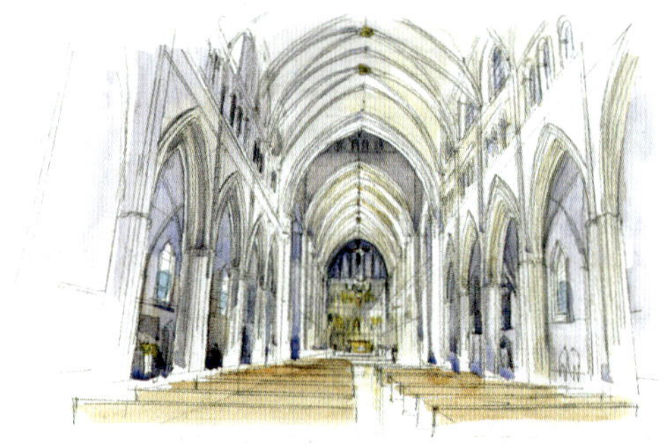

Southwark Cathedral, London, UK

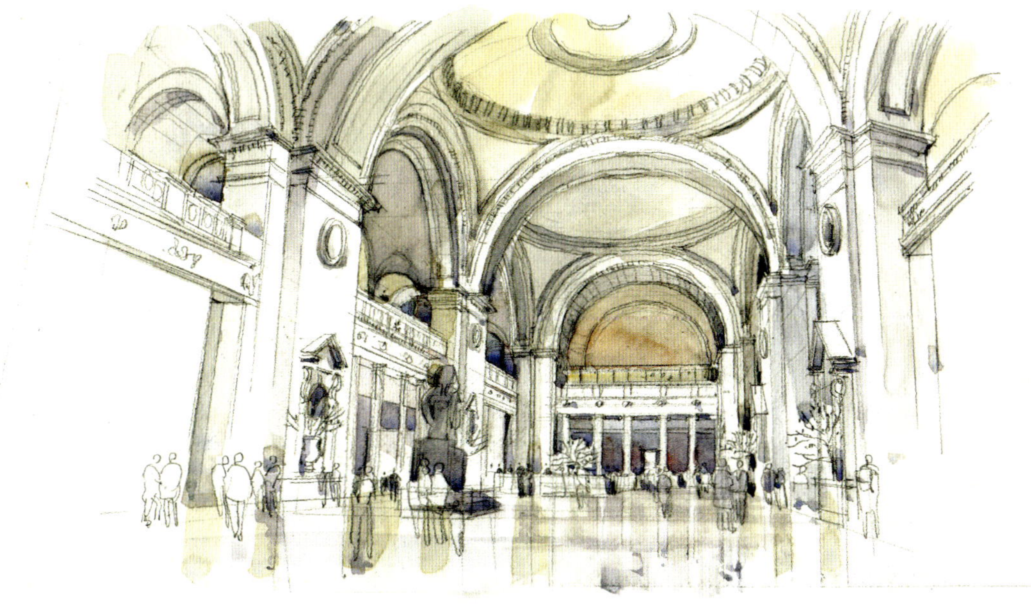

The Metropolitan Museum of Art, New York, USA

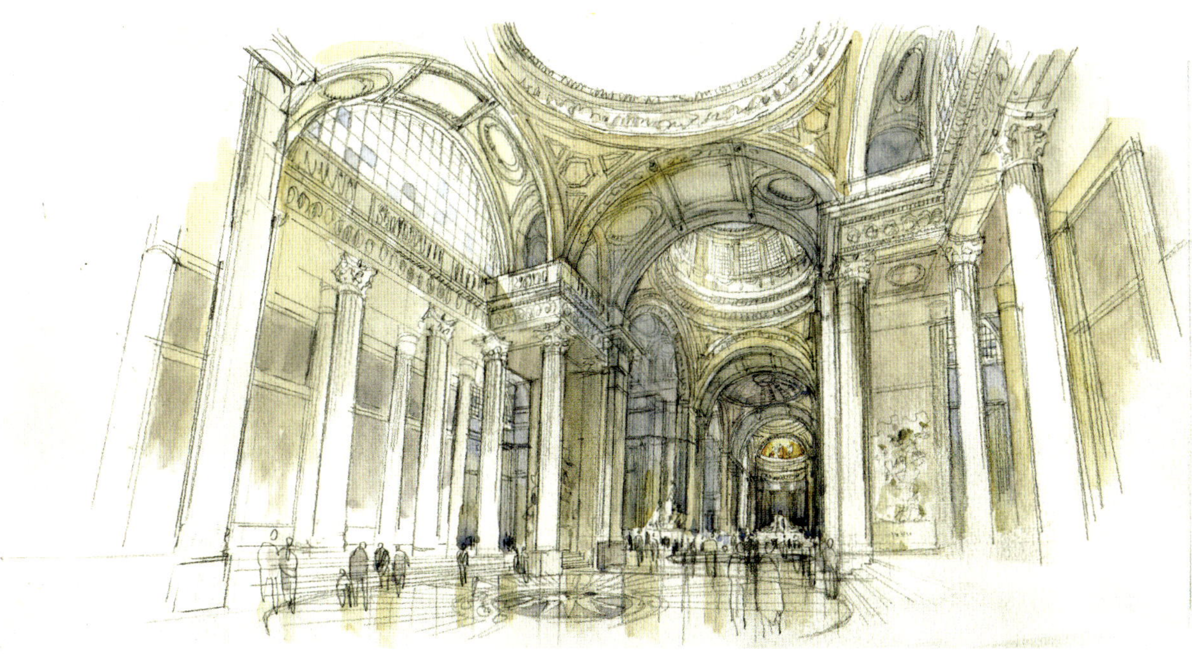

The Pantheon, Paris, France

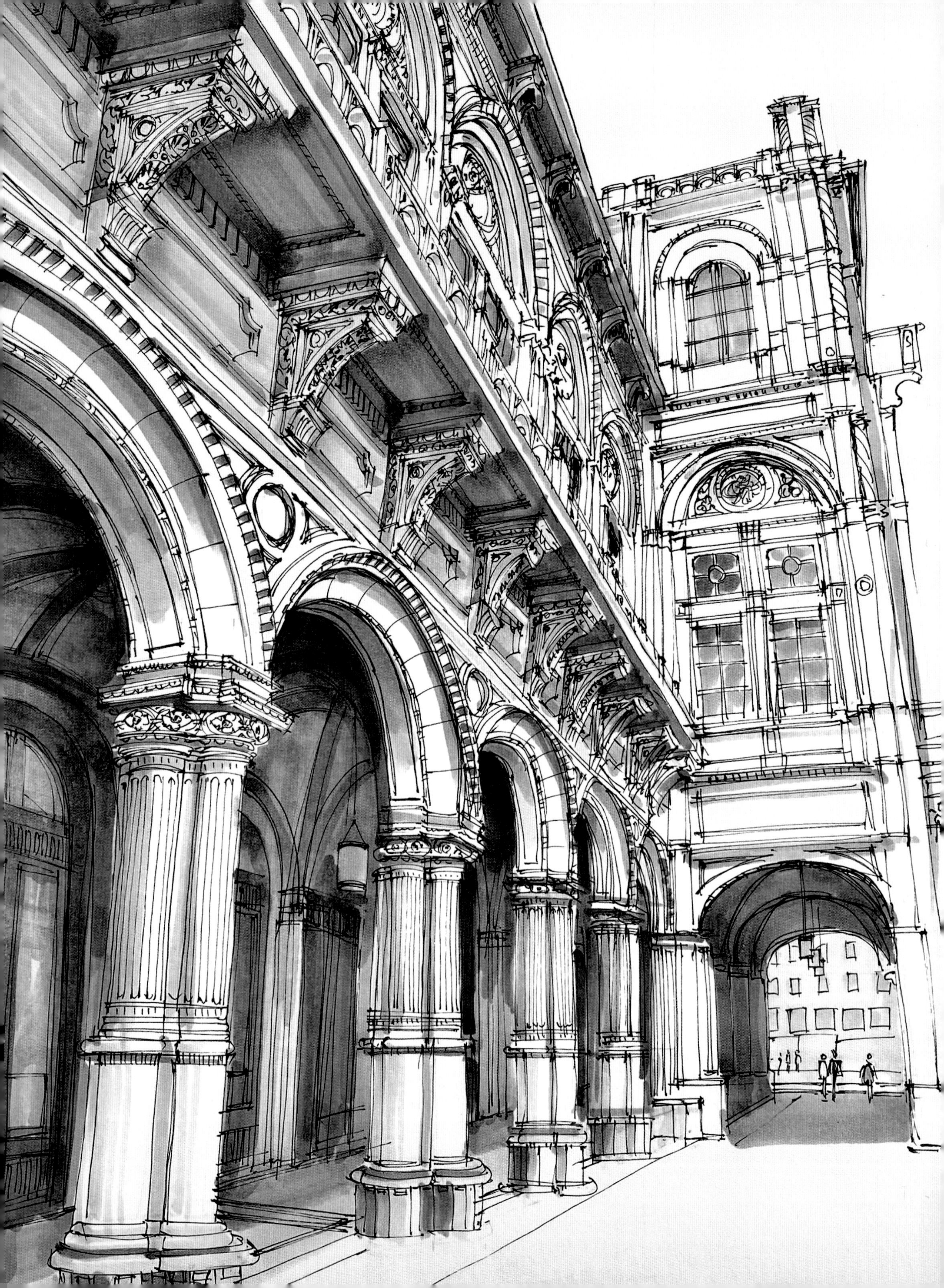

06

ADVANCED TECHNIQUES

In the following chapter, we'll look more closely at the advanced techniques of perspective drawing. Many of the examples included here draw on traditional draughtsmanship – techniques that would once have been taught as standard if you were studying architecture or interior design, before digital software became the industry default.

While they may feel a touch old-fashioned today, these techniques remain valuable, and in many cases the traditional ways still prove to be the best foundation for clear, precise drawing. I firmly believe that a regular drawing practice is essential for those studying or working in architecture or the creative industries. Hand drawing remains highly valued by clients for its immediacy and communicative power.

It is important that this final "advanced" chapter also acknowledges the role of digital software. Today's digital tools are highly sophisticated and there will be a digital solution to every perspective challenge discussed in this book. I also introduce drawing to architectural scale and the use of one- and two-point perspective grids, which are especially helpful when drawing interior spaces. Stairs, however, remain one of the most demanding perspective problems, bringing together many of the principles covered in this book. In fact, if you can confidently draw stairs in two-point perspective, I know my job here is done!

◀

A bold and confident partial side view of the Vienna Opera House, Austria, by Stephen Travers.

PERSPECTIVE GRIDS

There are various grids that can help with perspective drawing, allowing you to draw different viewpoints at the correct angle and to scale.

In this section, we will first look at using digital software to create grids. This is useful for speed and for making quick adjustments, but you need to be familiar with the software to use these grids effectively. We will then work through how to draw one-point and two-point perspective grids by hand, as it is useful to understand how they are constructed.

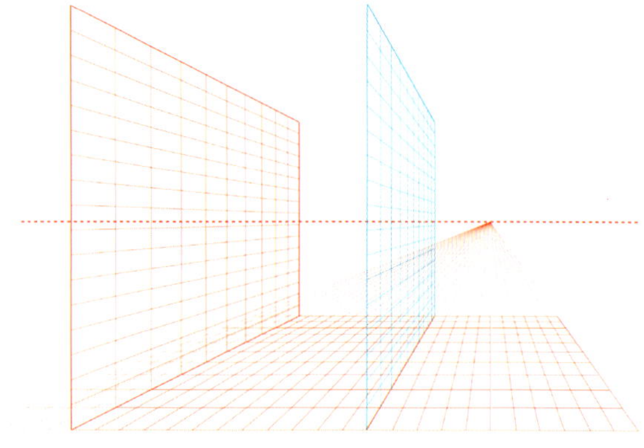

An example of a one-point perspective grid created in Adobe Illustrator

DIGITAL SOFTWARE

Programs such as Adobe Photoshop, Adobe Illustrator, and other digital design software provide effective tools for setting up perspective accurately.

The two grids shown here were created using Adobe Illustrator using the Perspective Grid Tool. The Perspective Selection Tool in this software allows you to adjust the angle of the perspective and move objects along the grid while maintaining accurate proportions. Once you gain confidence with the software, these tools become quite intuitive to use.

Procreate has become an especially influential digital drawing software, though it is currently available only on Apple iPads. It is useful for architectural drawing: its Perspective Guide lets you overlay one- to three-point grids, adjustable in colour, opacity, and line weight, while Drawing Assist snaps lines to the grid for accurate yet fluid sketches.

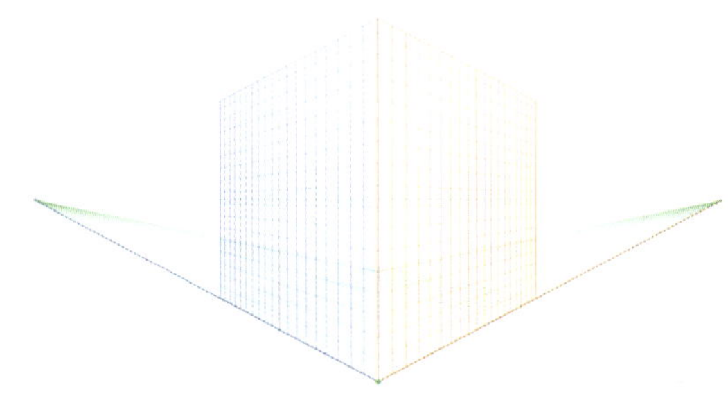

An example of a two-point perspective grid created in Adobe Illustrator

When overlaid on a photograph – with the centre aligned to the vanishing point – a grid provides a clear framework of receding lines, ensuring the drawn perspective is accurate.

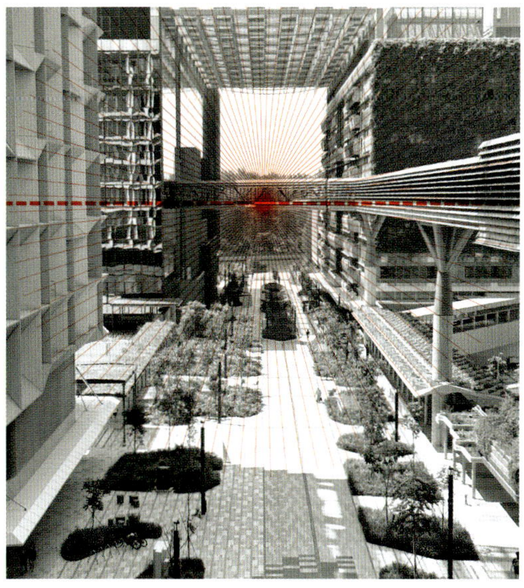

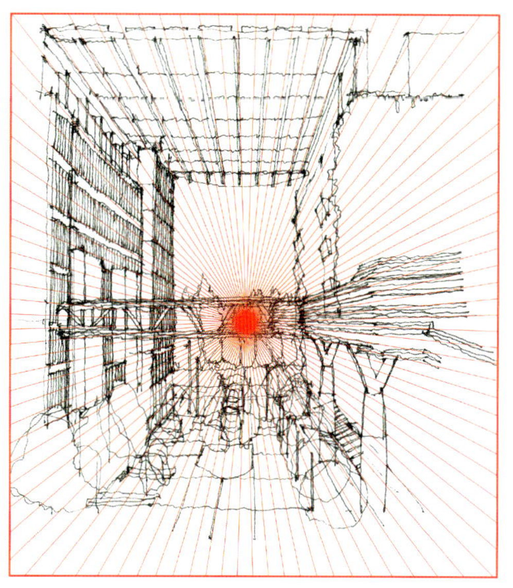

In this example, I've overlaid a Photoshop grid on a photograph and a sketch of the Singapore Institute of Technology. It's a clear demonstration of one-point perspective: the overlay shows how all elements recede towards a high vanishing point, both above and below eye level. You can clearly see how the grid defines the receding lines in the sketch. Although the sketch was drawn intuitively, the perspective remains remarkably accurate.

Tip

You can also use Photoshop – or other digital software – to create images in perspective. In this sketch of the Radisson RED Hotel at Lime Street Chambers in Liverpool, UK, I used the Perspective Tool to transform a front-facing elevation (left) into a perspective view (right).

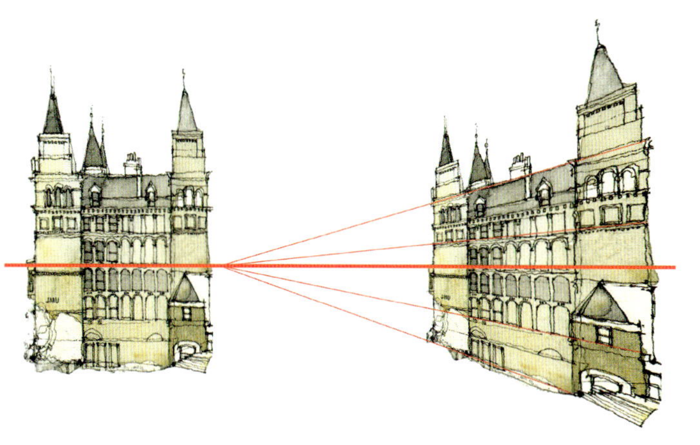

HAND-DRAWN GRIDS

One-Point Perspective Grid

Grids are particularly useful for interior design visualizations. Here is an example of a one-point perspective grid. Because the grid can be drawn to scale, it's especially helpful for working out the size of windows, doors, and furniture. It also makes it easier to measure equal divisions into depth, such as wall panels or floor tiles.

1. In the centre of your page, draw a rectangle twice as wide as it is high – for example, 120 × 60mm (4¾ × 2⅜in). Then divide this rectangle into a grid of 10mm² (⅜in²) squares – this represents the back wall. Add in the eye-level line (I've shown it at mid-height), then choose a vanishing point location on this eye-level line. In my example, it is placed between the fourth and fifth vertical grid lines (counting from the left).

2. From this vanishing point, project lines outwards to establish the side walls, floor, and optionally the ceiling.

3. To create a foreshortened floor grid, extend the eye-level line to the left. There are no fixed rules for how far to extend it, but here I have extended by approximately 150mm (6in) from the vanishing point. Next, draw a line back from this point through the lower-left corner of the back wall, ensuring it passes through this corner. Where this line intersects each division of the floor grid, mark a point.

4. From these points, draw horizontal lines until they meet the walls. This establishes the floor grid in perspective. You can then draw vertical lines to create a perspective grid for the side walls. With this grid in place, you now have a framework for accurately positioning interior elements.

1-2.

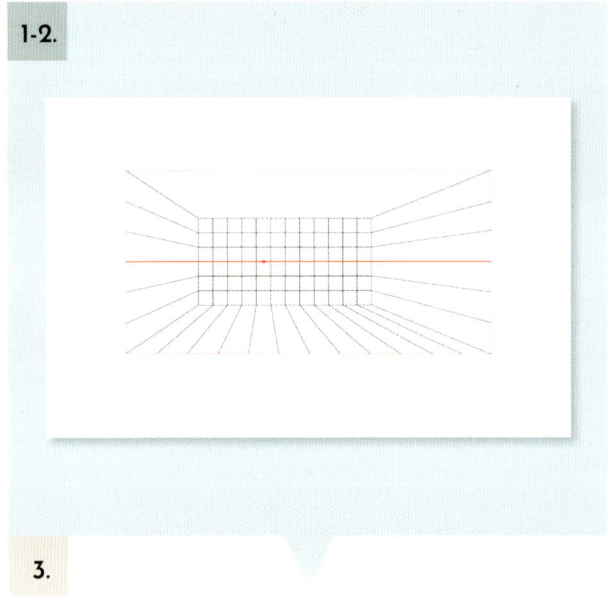

3.

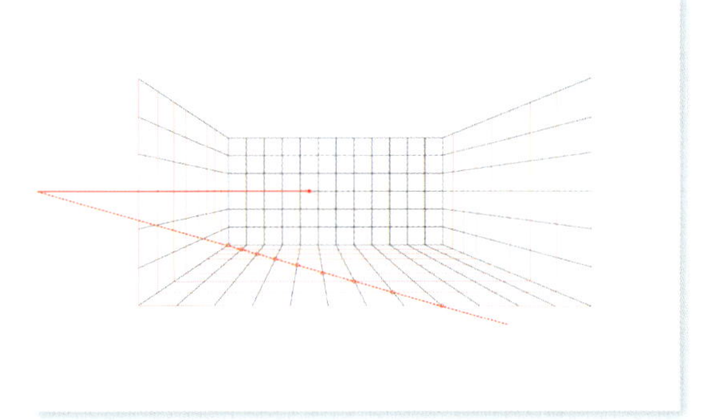

4.

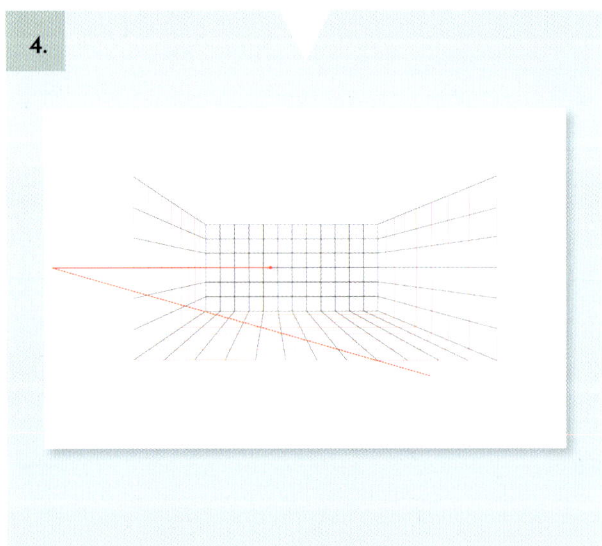

In this more developed drawing, I've added a door and two windows of equal size. As you can see, it's a more professional development of the simpler version in Chapter 02 – essentially the same view but now underpinned by the geometry that makes the perspective work. This is what we call a scaled grid, and this has been set at 500mm² (19½in²) per square, which makes the back wall 6000mm (236¼in) wide by 3000mm* (118⅛in) high. The eye level is placed at 1500mm (59in), while the door height is 2200mm (86½in). The windows are drawn to the same proportions, measuring three units high by five units wide. Notice how the grid ensures that the foreshortened windows are scaled correctly in perspective.

I've given the windows a notional wall depth. Because of the position of the vanishing point, the window facing us shows wall only on the right-hand side, receding at a shallow angle towards the vanishing point. For the window drawn in perspective, the wall depth is shown in elevation rather than perspective.

You can also experiment with how you position the diagonal line that generates the floor grid. If it's too close to the back wall, the grid will appear exaggerated and distort the foreground. If it's too far away, the grid becomes too shallow.

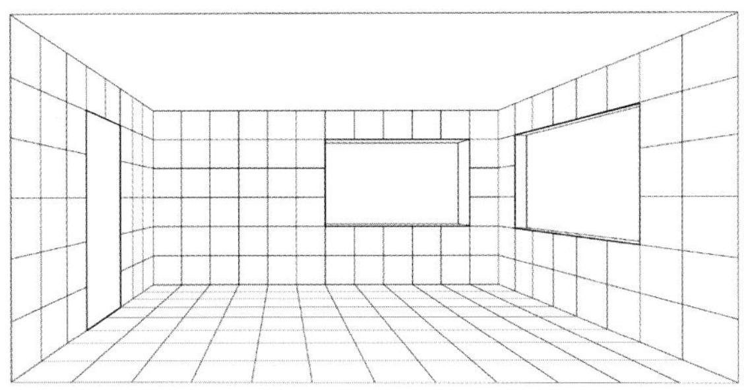

A more developed version of a one-point perspective interior grid

WORKING AT SCALE

Architectural drawings – plans, sections, and elevations – are conventionally drawn to scale because real objects are too large to fit on paper at full size. In the UK, common scales are 1:100, 1:50, 1:20, 1:10, and 1:5. A scale of 1:1 means actual size. At 1:100 (often used for building plans), 10mm (⅜in) on the page equals 1000mm (39⅜in) in reality. A scale ruler helps us measure and convert these proportions. Smaller scales are used for more detail: for example, 1:20 for kitchens or furniture layouts, and 1:10 or 1:5 for joinery or interior details. It's worth noting that when using digital software we usually draw at 1:1, since we're not limited by paper size. Everything can be created at its actual dimensions. Scale only becomes important when exporting or printing the drawing, as that's when we choose a scale (such as 1:50 or 1:100) to fit the design onto a sheet of paper.

Two-Point Perspective Grid

You can also draw a two-point grid in a similar way to the way we drew a one-point grid previously.

1. Start by drawing a vertical line 3500mm (138in) high in the centre of the page. This represents the internal corner of the room. Draw a horizontal eye level line at 1500mm (59in). Next, draw two vertical lines, each 4500mm (177in) from the central line, one on each side. Join the top of the central line to the top of each outer line, and repeat at the bottom. This defines the two side walls of the space in two-point perspective. Finally, from the points where these lines meet the outer verticals, draw lines back to the vanishing points. This begins to form the floor and ceiling grid.

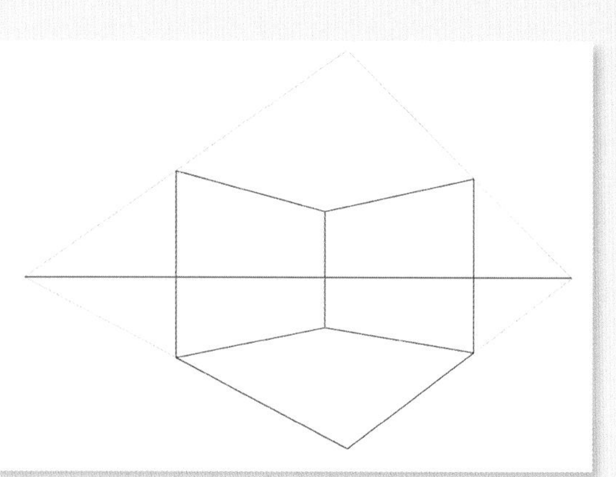

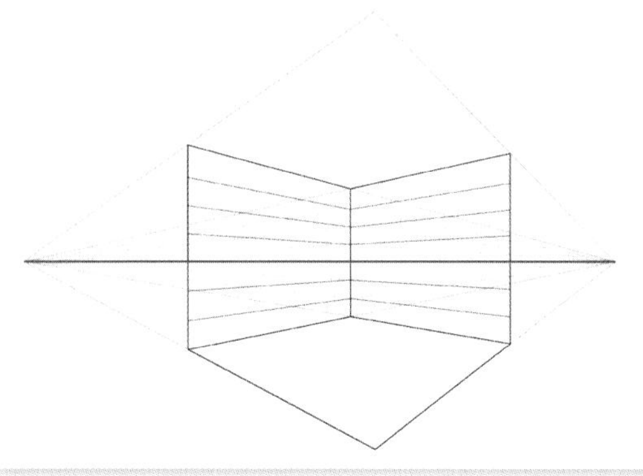

2. Divide the central vertical line into seven equal increments of 500mm (19½in): three below the eye level line and four above. From each of these points, draw lines to both vanishing points, extending them until they meet the outer wall lines.

3. Draw a diagonal line from the top of the left-hand outer vertical line to the bottom of the central vertical line. Repeat this on the right-hand side. Where these diagonals cross the receding lines, draw vertical lines. This creates the grid and ensures the perspective spacing is accurate.

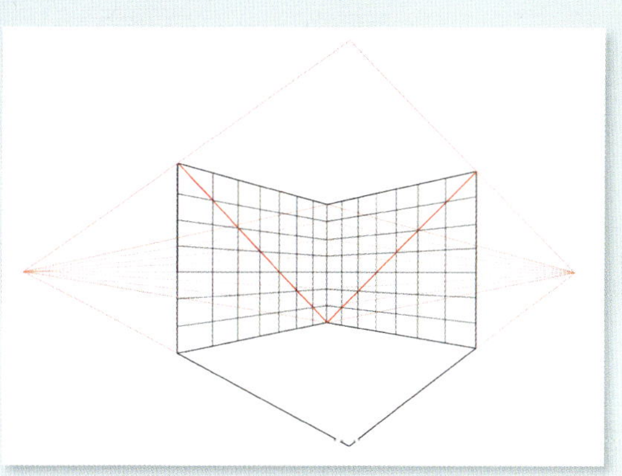

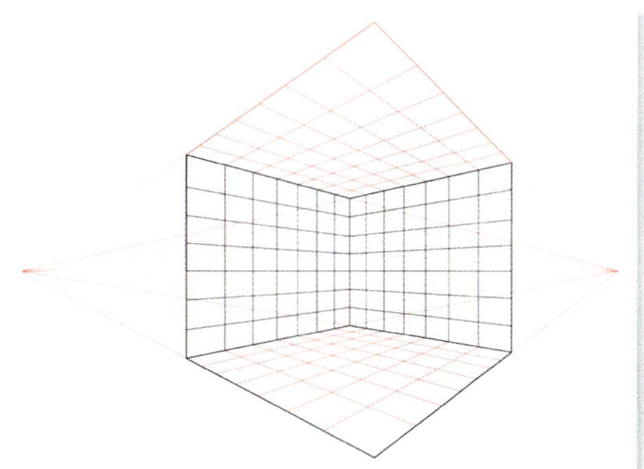

4. The perspective grid is now complete. If you like you can also fill in the floor and ceiling tiles as shown here.

Note: I set the vertical guidelines at 4500mm (177in) to create squares in perspective (which appear slightly distorted in the foreground), but you can experiment with different widths.

Tip

If one set of edges is directly facing you (not angled), a one-point grid is the right choice. If you're looking at an object from an angle, that's when you switch to two-point perspective.

SIMPLE BUILDING IN THREE-POINT PERSPECTIVE

1. First, establish your eye level. Draw a horizontal line across your page. Then draw a vertical line somewhere in the middle of your page – I suggest you make this slightly off-centre, to make for a more interesting drawing. Make this vertical line 25% below the eye level and 75% above. This will be the leading corner of your two-point perspective drawing. Then put two vanishing points on the eye level at either edge of the page and draw receding lines from the top and bottom of the vertical line. Now decide where to place the other external corners of the building (the corners we can see), these will be represented by vertical lines too. You now have a basic building form drawn in two-point perspective.

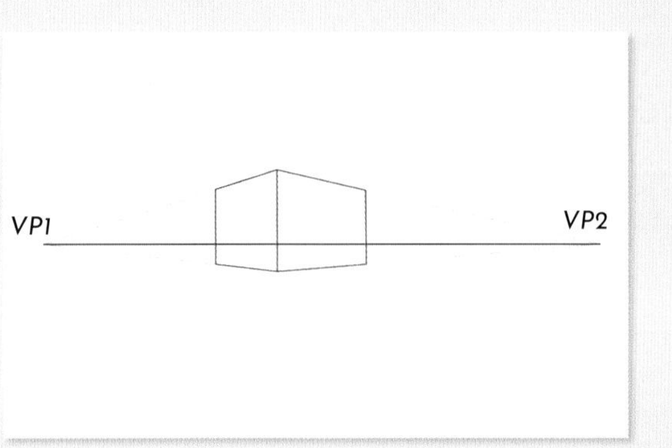

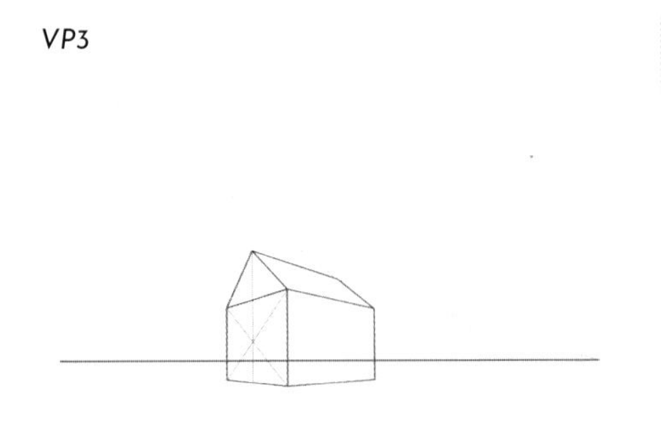

2. Next, you are going to put a pitched roof on the building. We want the pitch to be centrally located so we use the diagonal technique. From where these lines intersect, draw a vertical line and we can decide how high we want the roof pitch to be. The roof line will be parallel to the wall below so will recede to the same vanishing point (VP2), but the roof also slopes upwards as well as going away from the viewer. This means it doesn't recede to the left or right vanishing points, but towards a new vanishing point above, which shows the roof rising into space. This third vanishing point (VP3) is located directly above VP1.

3. Next, divide the building into notional ground and first floors. This allows you to begin adding doors and windows, using receding guidelines from VP1 and VP2 to guide their placement. Establishing the two storey levels helps make the building feel more convincing. You can also use the diagonal method to accurately position a central front door, then arrange the windows around it on both the ground and first floors. While the top and bottom of the windows are aligned using perspective guidelines – and we have already explored how to do this in one-point perspective – their horizontal spacing in this example is estimated by eye.

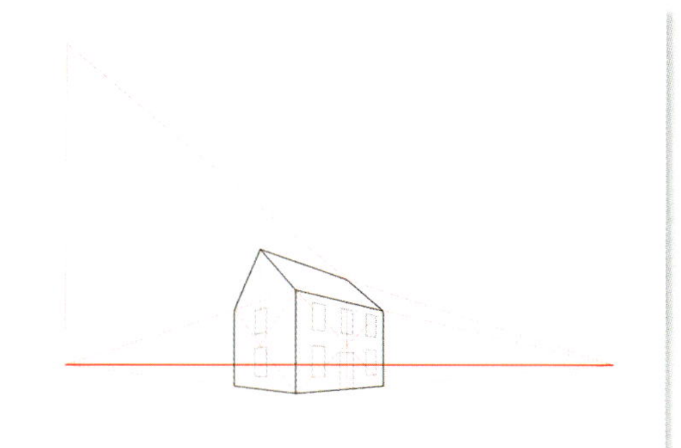

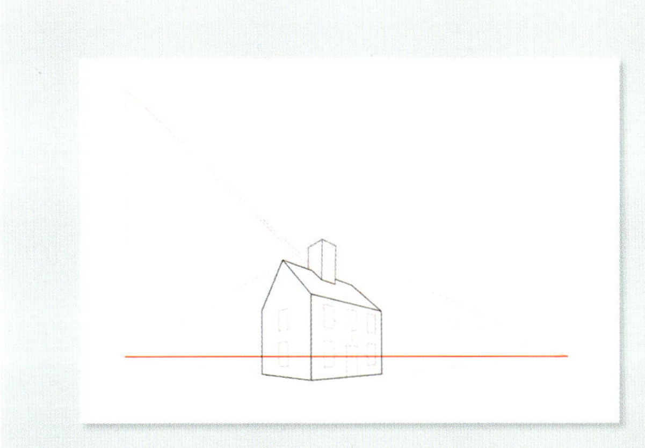

4. Then you can add a chimney, which in this example is positioned centrally, and as with the main building, you can establish a leading corner from which the chimney recedes in two directions to VP1 and VP2. The angle of where the vertical chimney sits on the pitched roof will be determined by a receding guideline back to VP3.

5. Finally, you can use your imagination to add landscape elements that also recede towards VP1 and VP2. These might include paths, roads, fences, trees, lamp posts, and of course, other buildings. Once these basics are in place, you can apply the same principles to a real-life building drawn in two-point perspective. Keep in mind that not everything will strictly follow the perspective rules – especially when sketching on location. Landscape elements often have a more irregular, unpredictable character!

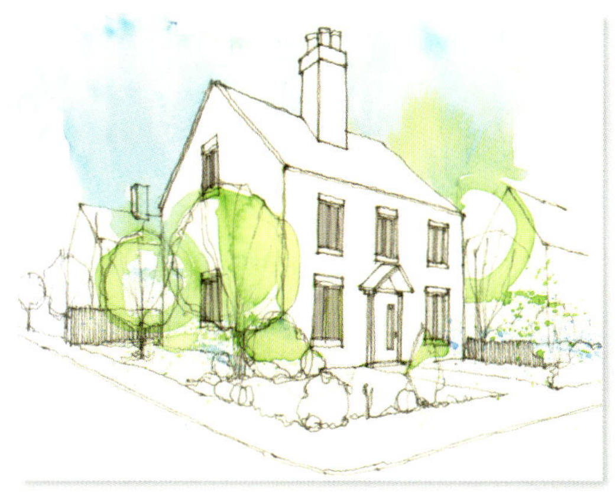

STAIRS IN ONE-POINT PERSPECTIVE

Staircases and changes in level offer an interesting and effective subject for practising perspective drawing. Let's begin by looking at a simple staircase in one-point perspective.

To set a staircase up in one-point perspective, start by locating the eye level. If we visualize the stairs as a three-dimensional form, the "footprint" of the staircase will recede towards the vanishing point (VP1) as you would usually expect (see diagram).

However, stairs don't just recede – they also ascend. If we closely observe a staircase, we'll notice that in addition to receding towards VP1, the steps themselves lead to a second vanishing point (VP2), positioned above VP1 but on the same vertical plane.

Another important detail is how the individual steps appear. Below the eye level, the horizontal surfaces of each step, known in architectural terminology as "treads", are visible. But above the eye level, only the vertical sides – called "risers" – can be seen.

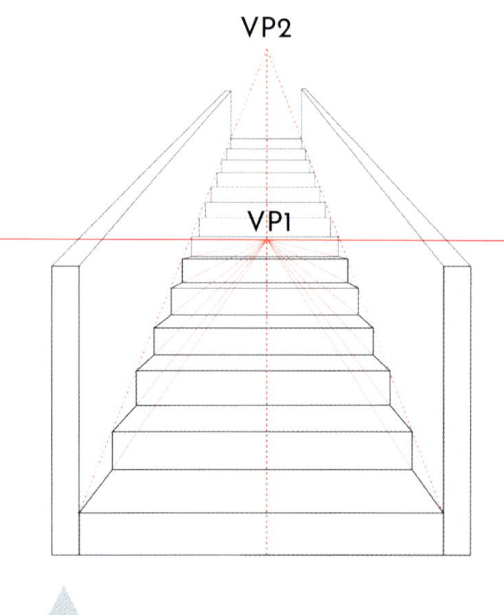

Diagram to show ascending staircase in one-point perspective

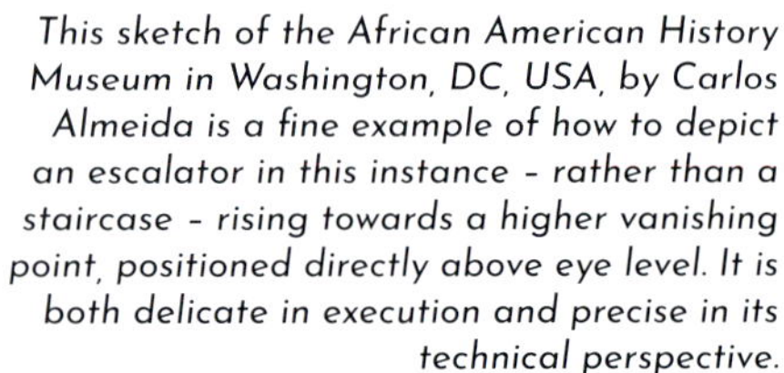

This sketch of the African American History Museum in Washington, DC, USA, by Carlos Almeida is a fine example of how to depict an escalator in this instance – rather than a staircase – rising towards a higher vanishing point, positioned directly above eye level. It is both delicate in execution and precise in its technical perspective.

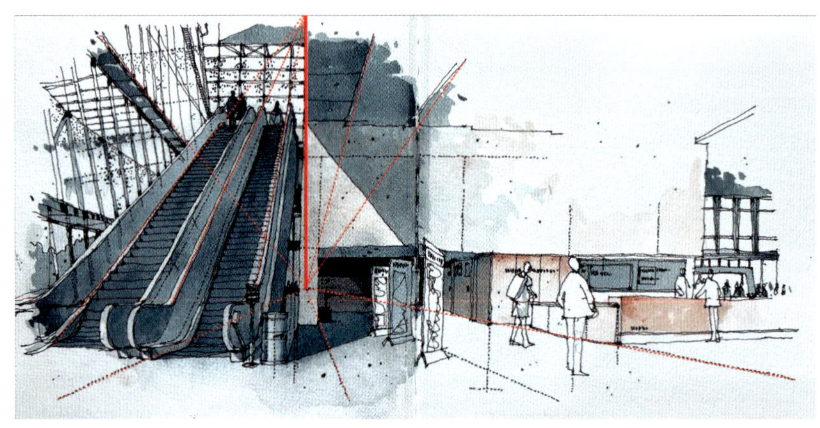

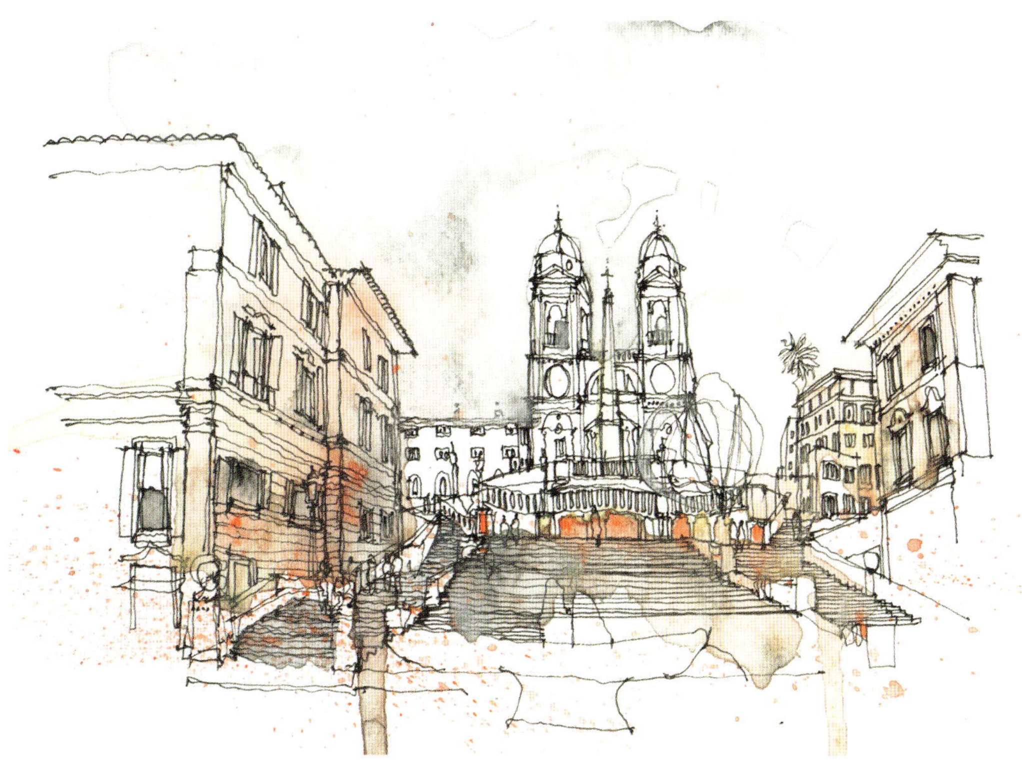

In this sketch, also of the Spanish Steps in Rome, Italy (see Darman Angir's sketch at the start of Chapter 05), the steps recede to a high-level vanishing point that is above that of the adjacent buildings, which recede to a much lower eye-level vanishing point.

EXERCISE

The same principles apply to escalators as well as staircases. Can you identify where the two vanishing points (VP1 and VP2) are in this photo of St Pancras International, London, UK? Remember that VP1 sits on eye level and defines the one-point perspective view of the entire space, not just the escalator.

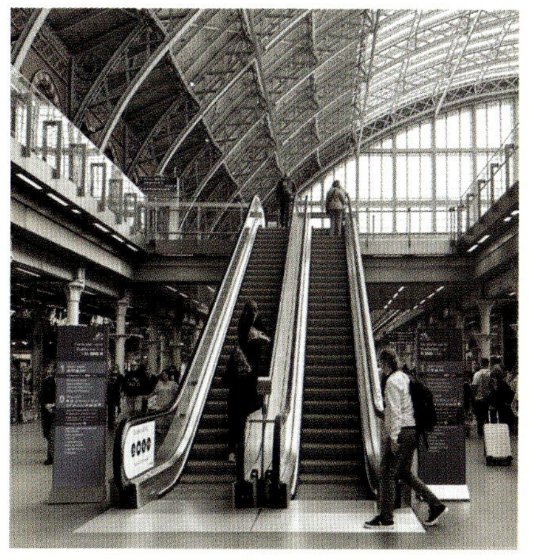

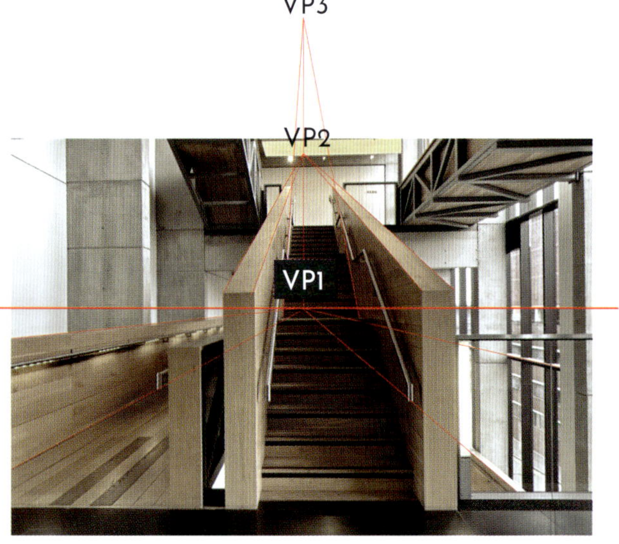

A staircase in Manchester Metropolitan University's Lowry Building, UK

This first photograph is a close-up of a staircase in Manchester Metropolitan University's Lowry Building illustrating the principle mentioned earlier, where an additional vanishing point (VP2) appears above the main vanishing point (VP1), but still on the same vertical plane. This is again a clear example of one-point perspective, with most elements receding towards VP1, except for the staircase. However, it's not quite that straightforward. The staircase includes two landings, which affect the geometry and shift the positioning of the vanishing points. The timber-lined balustrade of the staircase recedes towards VP2. The staircase itself has three distinct sections, with landings between them, and each section recedes to a different vanishing point from the main balustrade – though all remain aligned on the same vertical plane. However, in this photograph, it's difficult to clearly distinguish all three sections, and to avoid over-complicating the image I've indicated only the vanishing point for the highlighted middle section (VP3).

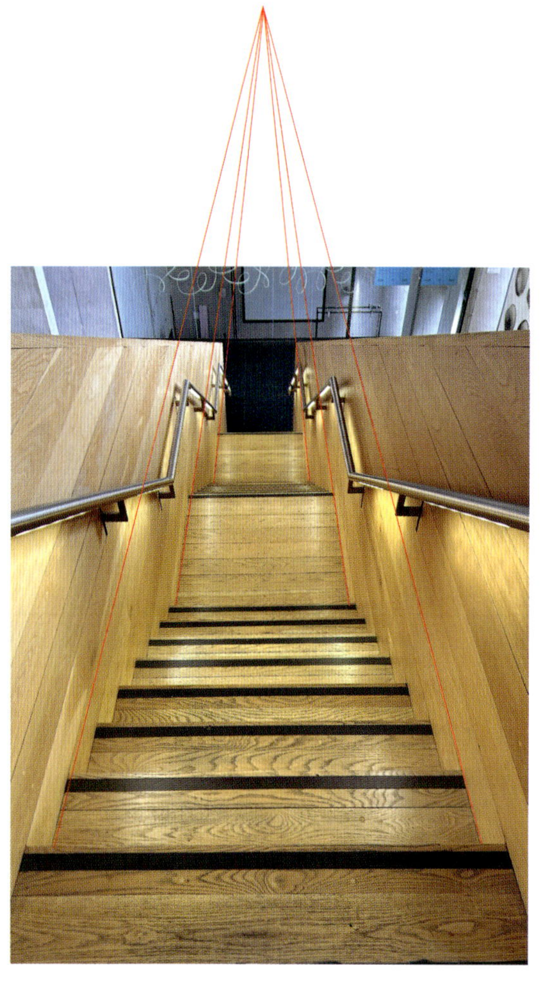

A different staircase, also in Manchester Metropolitan University's Lowry Building, UK

This photograph is another close-up of a staircase in the same building, but this time we're looking *down* towards where it meets the floor below. This view more clearly shows how the different sections of the staircase all recede towards the same vanishing point, while the balustrade recedes towards a different vanishing point (not marked in the image). Although the main vanishing point appears to be high above the staircase, this is technically a bird's-eye view – so in reality, this vanishing point sits *far below* the staircase. (It's a bit counter-intuitive, but it makes sense when you consider the perspective.)

STAIRS IN TWO-POINT PERSPECTIVE

A staircase in two-point perspective is another useful subject for practising perspective drawing.

Building on the underlying principles of understanding stairs in one-point perspective, we begin by establishing an eye level and setting our vanishing points – VP1 and VP2. VP1 defines the footprint of the staircase on the ground and VP2 controls the receding elements (treads and risers, as mentioned previously). However, to accurately depict the staircase's ascent, we must introduce an additional vanishing point (VP3) positioned above VP1, as we did in one-point perspective.

In this example, the same principles apply: below the eye level, the horizontal surfaces of the steps, known as treads, remain visible and, above the eye level, only the vertical sides, or risers, can be seen. This technique effectively captures the depth and elevation of a staircase within a two-point perspective framework.

For this example, I've drawn a 14-step staircase. The leading vertical line is divided into 14 increments to mark the steps. Eye level is set between steps 7 and 8 (about 1500mm (59in) high). From this point, guidelines are drawn back to two vanishing points, VP1 and VP2, which are spaced widely apart. A third vanishing point, VP3, is placed above VP1.

The stair width is estimated: I used the edges of the bottom step to project guidelines back to VP3, which establishes the staircase pitch.

The leading vertical line is divided into 14 increments to mark the steps.

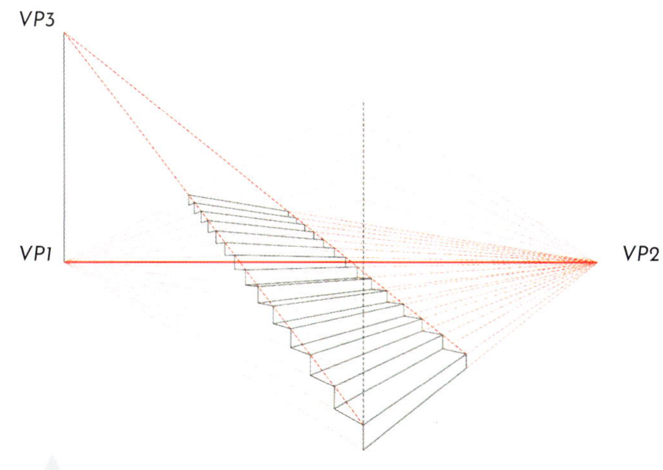

These receding lines provide the framework for the staircase, allowing the rest of the steps to be completed in perspective.

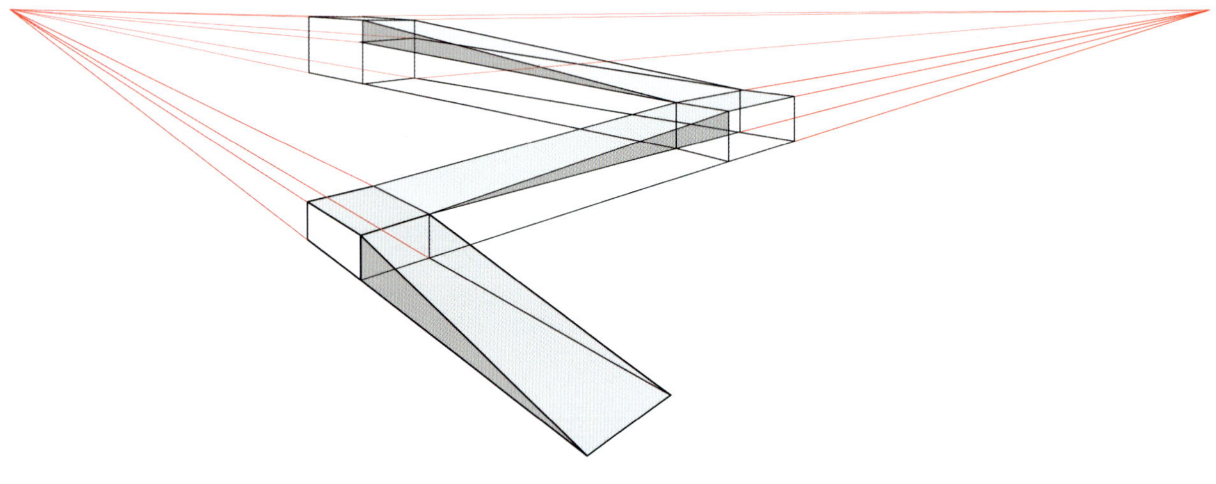

This diagram shows a ramp, made of three sections, each connected by a flat landing, drawn in two-point perspective. I've used two vanishing points to establish the overall view, and each section of the ramp rises incrementally by the same amount. However – and this is where it can be confusing – because the ramp is also receding in perspective, the upper sections appear shallower even though their actual incline is identical. The shaded areas on each section help make this clearer.

It's also important to understand how two-point perspective staircases function in real-life settings. This photograph is of the atrium at The Faculty of Science and Engineering at Manchester Metropolitan University. It's an interesting example: not only is it a staircase receding towards two vanishing points, but it's also situated adjacent to a bank of stepped seating. In interior design, we refer to this kind of built-in feature as "furnitecture". As you can see, the stepped seating is the height of two risers.

The staircase and seating both recede towards a right-hand vanishing point (VP2), located at eye level, while the external windows recede to a left-hand vanishing point (VP1), also on the same eye level.

As with staircases drawn in one-point perspective, there's an additional vanishing point (VP3) positioned vertically above VP1, accounting for the upward direction of the stairs.

However, a closer look reveals something more: there's a landing halfway along the staircase. This introduces yet another vanishing point – VP4 – to which the lower section of the stairs and seating appear to recede. Remember higher vanishing points are aligned on the same vertical plane above the eye-level vanishing point (VP1).

Honestly, I might not have noticed this just by looking. It was only through overlaying guidelines on the photograph that this became clear. It's a great reminder that real-life perspective is rarely straightforward – and careful observation often leads to new discoveries!

This overlay shows a staircase and stepped seating in the atrium of The Faculty of Science adn Engineering at Manchester Metropolitan University, Manchester, UK.

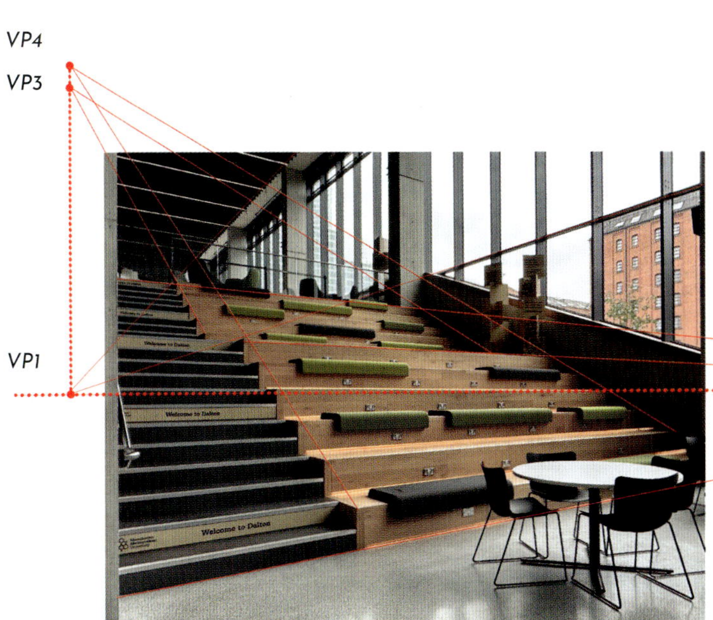

FEATURED ARTIST

Carlos Almeida

Baltimore, USA

The world is three-dimensional. Anything we see around us is visually perceived in three dimensions. The understanding of perspective is essential to represent objects, buildings, people, and nature. But it does not end there. There are a multitude of factors that need to be taken into consideration.

When you arrive at a certain place, we feel attached to it, and many aspects are determinant to capture the sense of the place. Some of these aspects are related with the point of view and angle, at a certain moment. The light at a certain time of the day, texture, contrasts, people, human-made urban elements, and many others – these are all variables in the entire equation that we want to solve and, in order to tell the story, are preponderant in any hand-sketch or urban sketching. It's all about the story.

Hand-sketches include all these elements but there is one that definitely translates the reality of the place: the perspective.

Perspective can be very complex with so many vanishing points, depending on the scene complexity.

If we don't rush to immediately draw a certain scene, and instead we take the time to understand and dialogue with it, this may be the right approach. From this point, there's a story that can be as dramatic as the selected perspective. Sometimes, it's the very first impression of a place that ought to be captured. That very first impression is normally the most truthful one, and unforgettable.

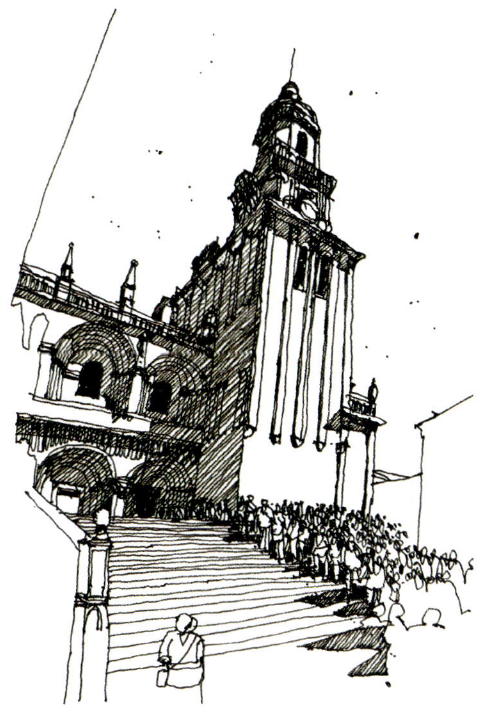

Cathedral of Santa Maria Compostela, Santiago, Spain

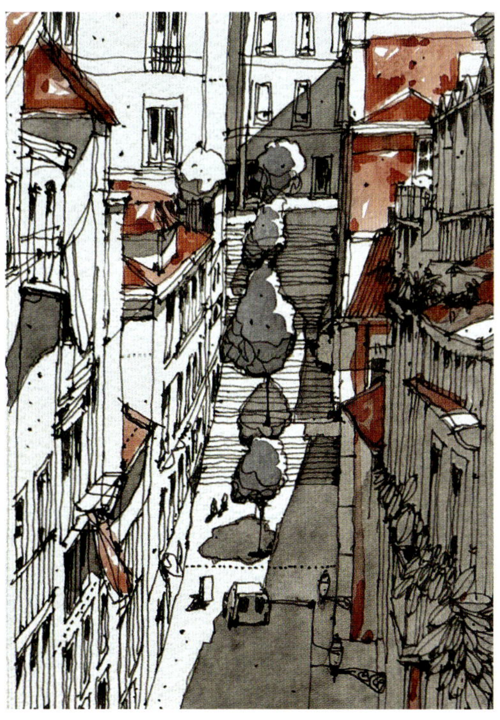

Steep steps, Lisbon, Portugal

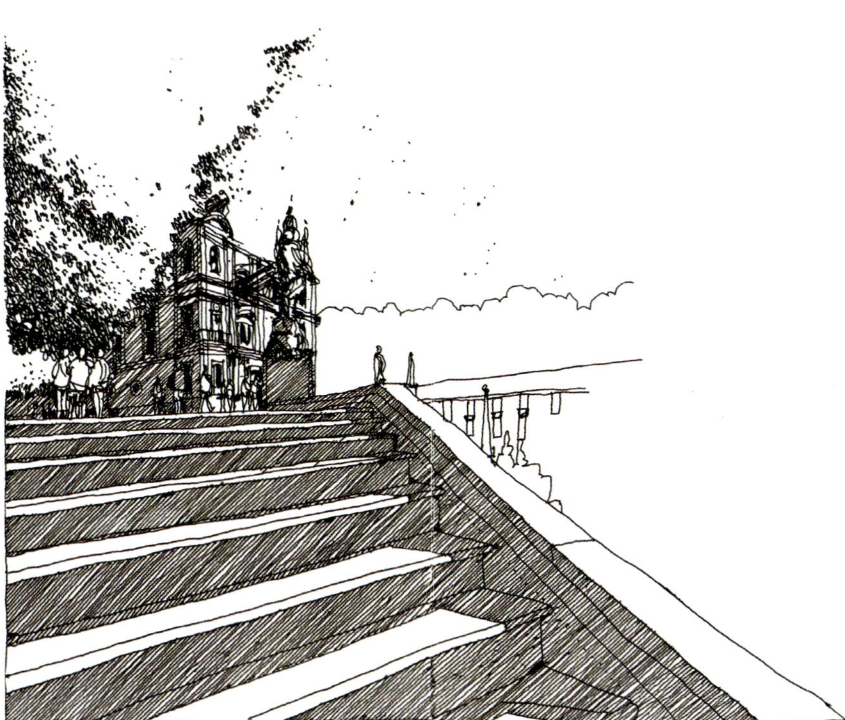

The Sanctuary of Bom Jesus do Monte, Braga, Portugal

INTERIORS

Often, as an architect or interior designer, you need to illustrate spaces that don't yet exist. Clients don't always understand standard architectural drawings such as plans and sections, which is where perspective drawings – or interior visualizations – become invaluable. You can represent an unbuilt or hypothetical space space with quick one- or two-point perspective sketches or set up a more technically accurate perspective view based on your plan and section information.

That said, many designers today go straight to digital tools, using 3D modelling software such as SketchUp or AutoCAD, or outsourcing to specialist visualization companies. Still, it's valuable to understand how to create perspective sketches manually – and clients often appreciate the immediacy and personal touch of freehand drawings!

This sketch by Stephanie Bower of the Graphite Arts Center in Edmonds, Washington, USA, convincingly captures an interior one-point perspective view. It was not constructed from a set-up plan but drawn directly – Stephanie's experience as both architect and illustrator allows her to instinctively depict the perspective. Although a few angled walls and a ceiling feature depart from strict one-point perspective, the drawing succeeds through her confident handling of tone and colour. Her trademark watercolour shadows and harmonious palette unify the composition. The feature ceiling lights introduce a sense of dynamism, while the subtle figures establish scale and bring the interior to life.

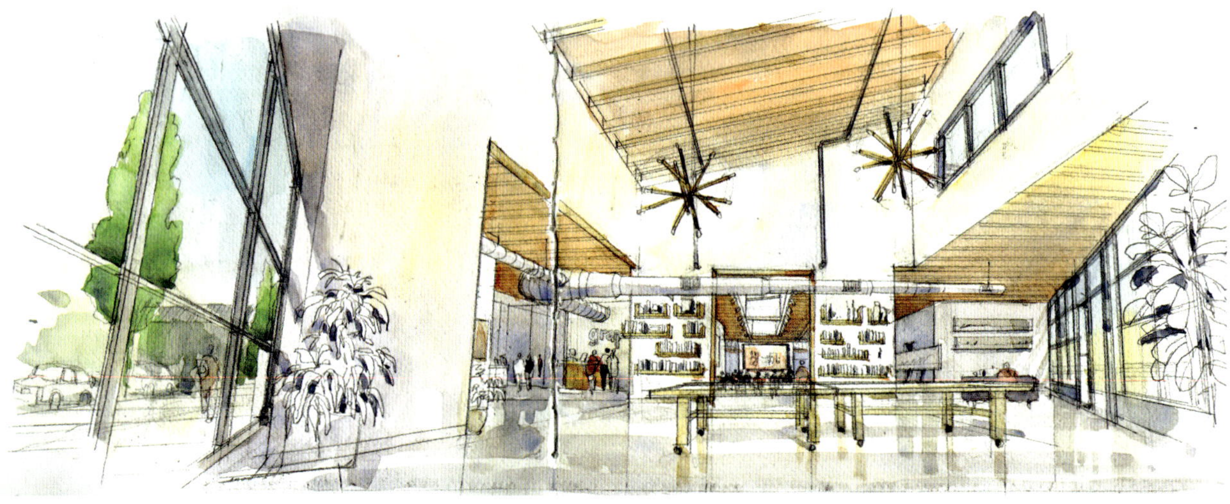

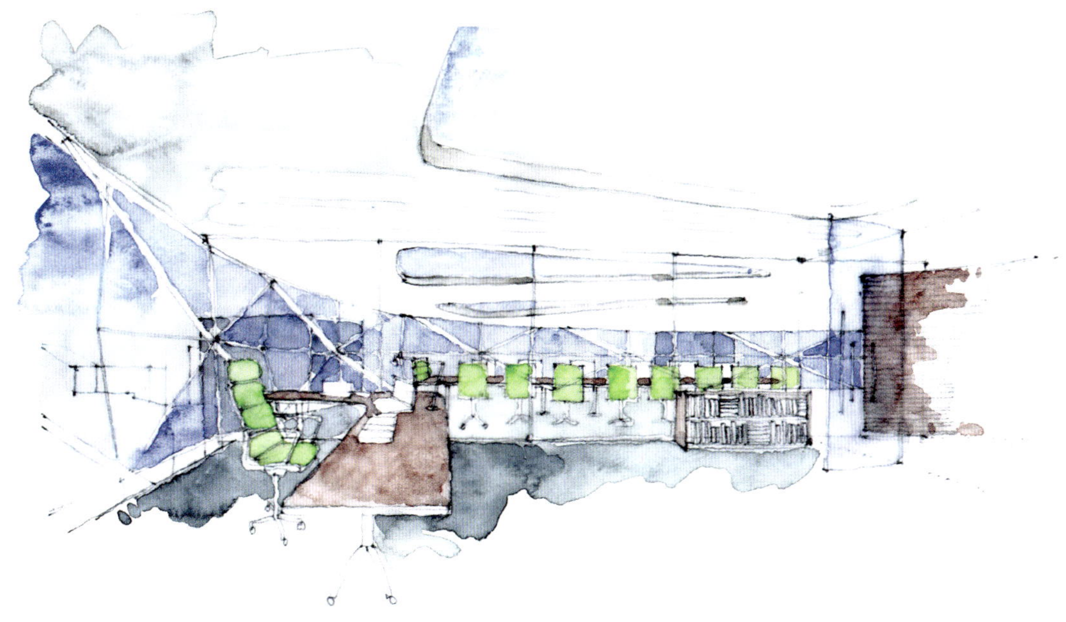

This loose one-point perspective interior sketch was created for a client presentation drawn in pencil and watercolour.

ROOM FROM A PLAN

Let's now look at setting up a one-point interior perspective. This is going to be a room with lounge furniture in it. As you can see, the plan shows a sofa, armchair, coffee table, dining table with two chairs, and and a sideboard set into the window alcove.

By creating a ground-floor grid which will recede to a vanishing point, we can establish the floor plane for objects and furniture within the interior space. From a defined measuring point, heights can then be set for the furniture in plan and extruded vertically to occupy the space.

In this example, as with other interior examples in this book, I have set eye level at 1500mm (59in). The room is 6000mm long × 3000mm wide (236¼in × 118in).

Plan vs Plan View

A plan is a technical drawing drawn from above showing walls, doors, and furniture, usually drawn to a recognizable scale such as 1:100.

A plan view, in perspective drawing, is the same top-down view but is used as a construction tool to set up a perspective drawing. It is typically used alongside an elevation and is needed to position the viewer (station point) and picture plane. The plan view acts as a set-up drawing to help construct the perspective.

1. Set up the grid:

a. Once you have a scaled plan drawn (1:50 is good for this exercise), then overlay a grid of 500mm (19½in) squares at the same scale as the plan.

b. You could use a different colour for the grid so it's clearly differentiated, and you may find it helpful to number the rows horizontally and vertically.

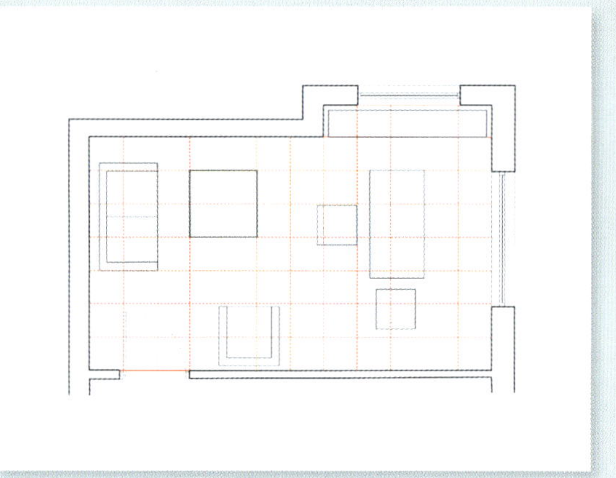

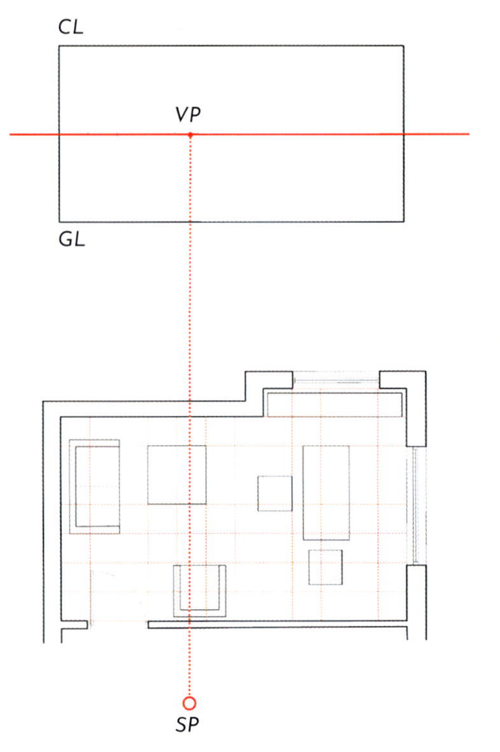

2. Establish your station point and eye level:

a. Decide where you want to stand in your perspective drawing and mark this as your station point (SP). Note that to give a more realistic perspective view, I have positioned this "outside" of the plan.

b. Overlay tracing paper onto your plan. Then draw a horizontal line for your eye level and place a vanishing point centrally on the eye-level line, in line with the station point.

c. From there, measure 1500mm (59in) down to mark the ground line (GL), and 1500mm (59in) up to mark the ceiling line (CL), assuming a 3000mm (118in) room height.

d. Connect these horizontal lines with verticals to define the inside edges of the plan. This becomes your back wall.

Note: I have shown the overlay above the plan drawing, as placing it directly on top would make it confusing to read. However, when you carry out this exercise yourself, the perspective view should be drawn on tracing paper placed over the plan.

3. Project the grid: using the same grid as the plan, mark 500mm (19½in) increments, project these vertically upward. As the height of the room is 3000mm (118in), you can draw horizontal lines at 500mm (19½in) increments to create a complete grid on the back wall.

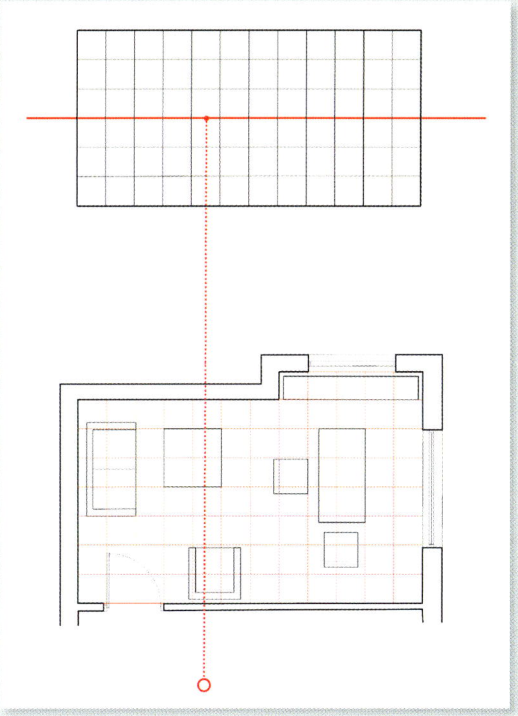

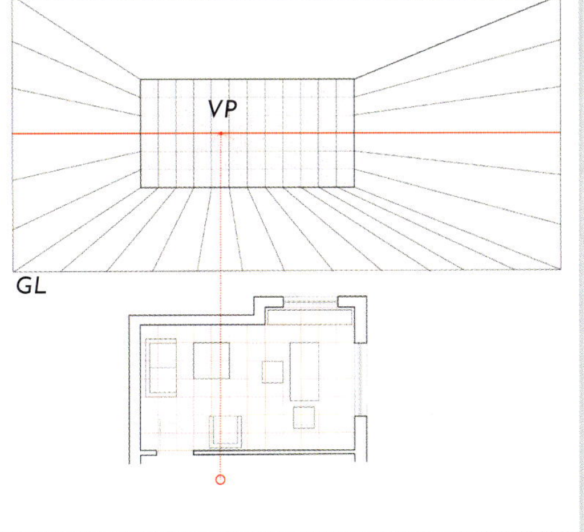

4. Extend the grid:

a. From your vanishing point (VP), draw lines through each corner of the back wall and extend them forward to the edge of the tracing paper sheet so the drawing looks similar to the diagram.

b. Repeat this from the VP through each 500mm (19½in) division on the GL to establish your floor grid.

c. Apply the same process to the side walls so the grid extends evenly in perspective. You'll notice the spacing between lines increases towards the edges of the page – an effect that occurs in linear perspective.

Note: The side walls are not precisely measured in perspective because no measuring system has been used, but they remain in the correct proportions as they are based on the plan grid.

5. Add the measuring point:

a. To show the floor grid in perspective, I have extended the eye-level line from the vanishing point by approximately 7500mm (24ft 7in), so that it sits beyond the left-hand wall. This establishes the measuring point (MP).

b. From the MP, draw a diagonal line through the bottom left-hand corner of the back wall and extend it across the floor.

c. Wherever this diagonal line cuts the projected floor lines, draw horizontals to complete the foreshortened grid.

d. Extend these lines up the walls to create a wireframe effect – walls and floor drawn in correct perspective.

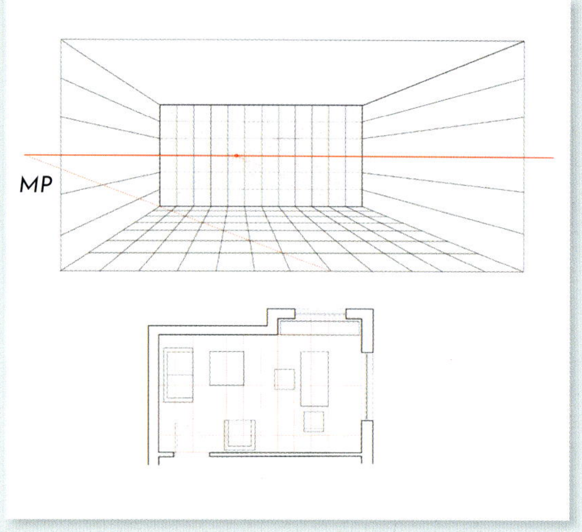

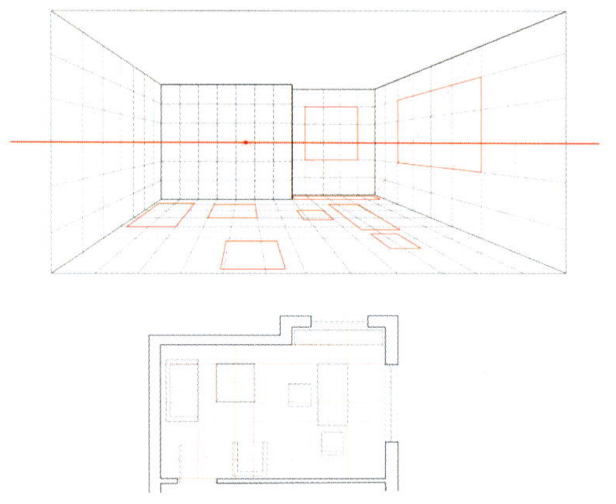

6. Plot furniture in plan:

a. Use this grid as a framework to plot furniture in plan view before elevating it.

b. Measure furniture by counting how many grid squares it occupies and transfer these proportions into perspective.

c. You can also plot in the windows.

d. Expect slight distortion near the edges of the page, consistent with the cone of vision. Objects in the foreground will obscure those further back.

Note: In this example, I've used diagonals to extend the floor plan into the alcove by one grid square. You can see how this was plotted using a diagonal construction line, shown as a red dashed line on the right-hand wall. If you prefer to create a simpler rectangular room, that is absolutely fine as well!

7. Extrude into three-dimensional forms:

a. Heights are always referenced from the back wall (drawn to scale).

b. Once satisfied that the plan positions match, begin extruding the furniture upward into three-dimensional blocks.

c. Keep the forms simple until proportions are correct.

Tip: Setting up a perspective grid is complex and sometimes the view may look a little distorted, which is due to the positioning of the side walls and the measuring point, so you can play around with this – but ensure it is on the eye level and outside the planes of the "back wall".

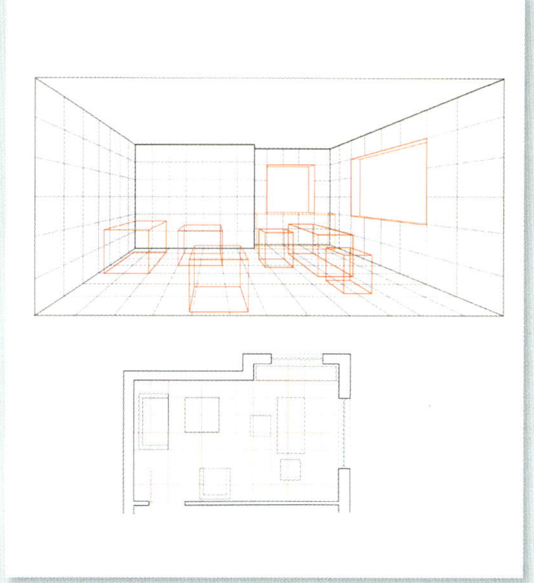

Finished one-point perspective visual showing furniture created from 3D forms extruded from the plan, with additional architectural details including windows and a fireplace.

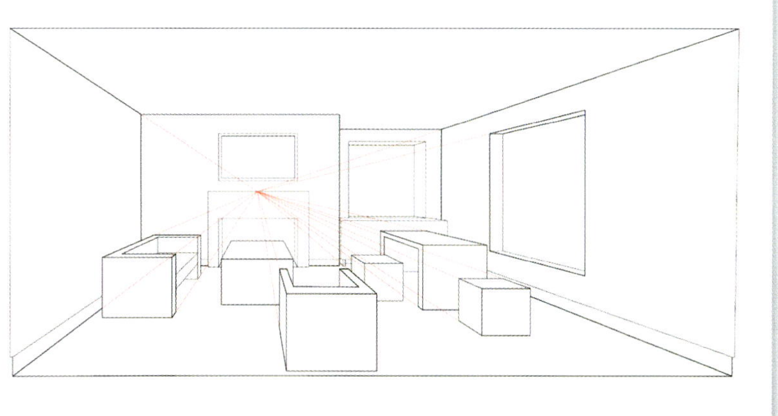

KITCHEN FROM A PLAN

Using the same principles as the drawing of a room from a plan, I've created a perspective view of a kitchen. I've used a plan and established a station point. This time, the kitchen plan has been simplified into a straightforward rectangular room.

As with the lounge example, the Station Point (SP) is placed slightly outside the plan to create a more natural perspective view. If it were positioned within the room, the view would appear more distorted and the kitchen units would look compressed.

The lower perspective view (above the plan) illustrates how the floor grid was established. A diagonal line was used to divide the left-hand wall into six equal panels, which in turn defined the grid for the floor. In this diagram, the elements highlighted in red indicate the components that were later extruded to form furniture in the final view.

Both the plan and perspective drawing have been produced to scale, with a ceiling height of 2400mm (94½in) and an eye level of 1500mm (59in). The horizontal lines on the back wall represent 600mm (23½in) increments, which assist in determining the height of the kitchen units, as these are typically designed in 600mm (23½in) module.

This example shows how valuable perspective set-ups can be for clients, especially when fixtures and fittings are drawn to scale.

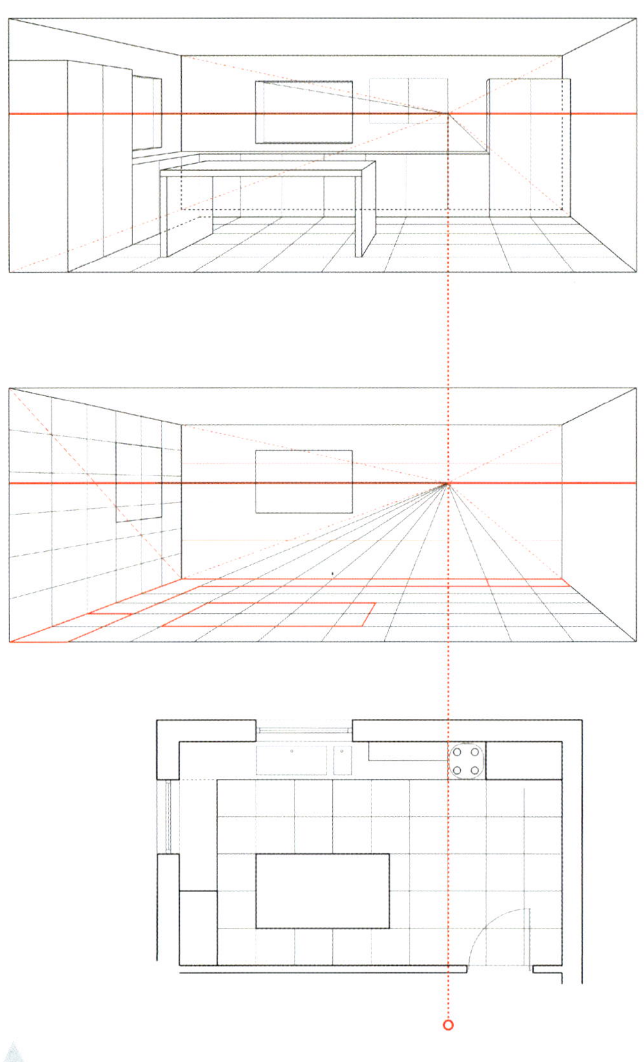

Perspective view and plan

Sectional Perspective

This is when a vanishing point is positioned within an architectural sectional drawing. This technique allows elements within the section to follow the rules of perspective, with walls and floors receding towards the vanishing point. Architects and designers favour this as a drawing type as it combines the clarity of a section with the depth and spatial understanding of a perspective view – offering a more comprehensive visualization of a building's interior.

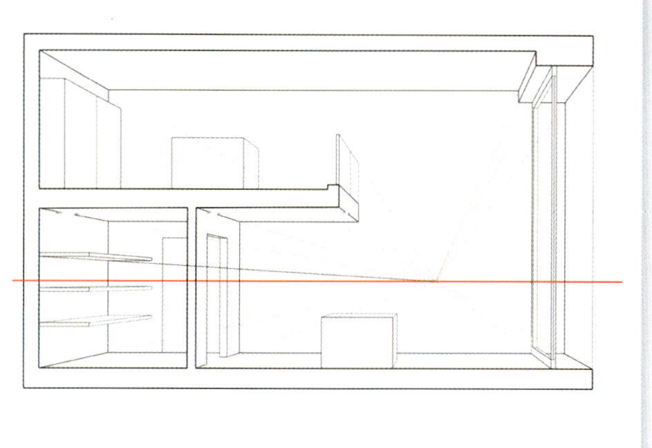

Sectional perspective showing an imagined double-height space, with the eye level and vanishing point set at ground-floor eye level.

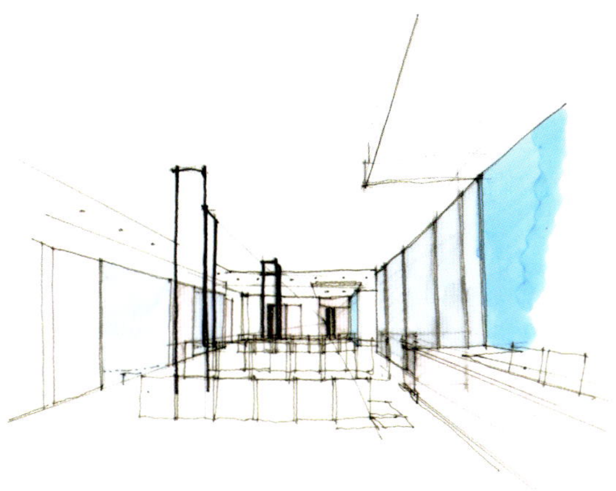

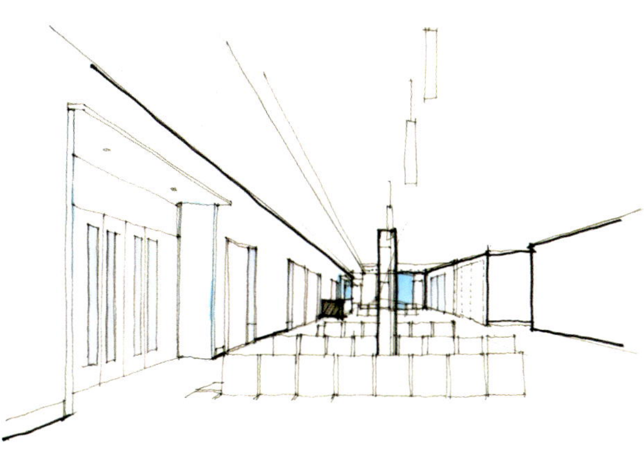

These drawings were created as quick concept visuals for a client presentation. They show loose one-point perspective views of waiting areas outside courtrooms, giving an approximate sense of the space and how it might appear in use.

OBJECT FROM A STATION POINT

The station point represents the viewer's position, and from this we'll construct a rectangular shape in two-point perspective using a simple plan and elevation.

1. Draw three equally spaced horizontal lines across your page: the top line is the picture plane, the middle line is eye level, and the bottom line is the ground line.

2. Draw the plan (top-down rectangle): draw a rectangle so that its bottom-left corner touches the picture plane. Rotate the rectangle so its sides are set at 60-degree and 30-degree angles to the picture plane.

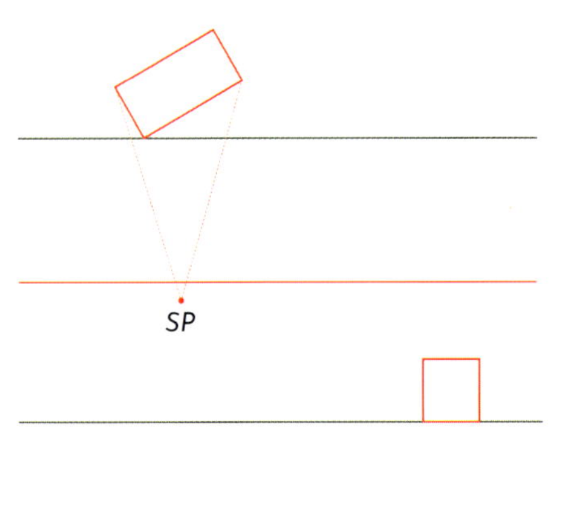

3. Place the station point (SP) somewhere near the centre of the page, below eye level. In this example, it sits just beneath eye level but well above the ground line. From the left and right corners – and from the corner of your rectangle that touches the picture plane – draw light projection lines back towards the station point.

4. On the ground line, add an elevation view, which is a front view; this will give you the height of your rectangle in perspective.

5. Find the vanishing points: from the station point, draw a line parallel to the left- and right-hand side of your rectangle. Where these lines intersect with the picture plane, drop vertical lines down to the eye-level line – these intersections are your two vanishing points (VP1 and VP2).

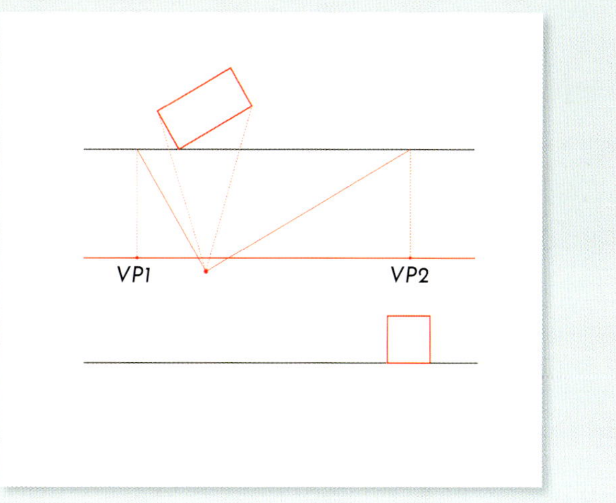

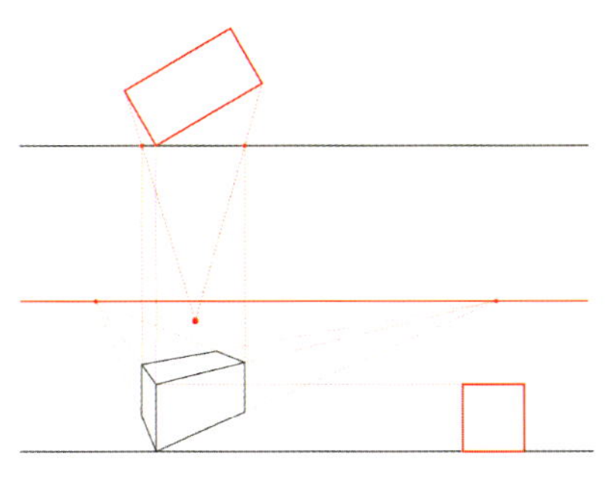

6. Project into perspective: from the bottom-left corner of your rectangle in plan (this is the one touching the picture plane), drop a vertical line down to the ground line. Then project a horizontal line across from the top of your elevation until it meets this vertical. This marks the leading corner of the object in perspective. From here, use the vanishing points to complete the rest of the rectangular form in two-point perspective.

Note

Remember, the station point represents the viewer's eye position, with the picture plane acting like a window between the viewer and the scene. Understanding this relationship is essential for setting up traditional architectural perspectives. Before digital software, this method was a fundamental tool for architects to create accurate visualizations.

HOUSE FROM A STATION POINT

This follows the same principles as drawing an object, but here we are constructing a single-storey building with a pitched roof. The plan is more detailed, including a door and windows. You need to have the plan and elevation of the building to be able to use this method of drawing a two-point perspective view.

1. Draw a horizontal line and mark this as the picture plane, then sketch the plan view (a top-down rectangle) so that its bottom-left corner touches the picture plane. Rotate the rectangle so its sides sit at 60-degree and 30-degree angles to the picture plane.

2. Below the picture plane, draw a ground line and add the elevation of your building along this line, to the right of the plan. Mark the station point (SP) – the viewer's position – somewhere near the middle of the page, below the ground line. Also draw the eye-level line. In this example, the house is drawn approximately to scale: the eaves are about 3000mm (118in) high, so the eye level is placed halfway up at 1500mm (59in).

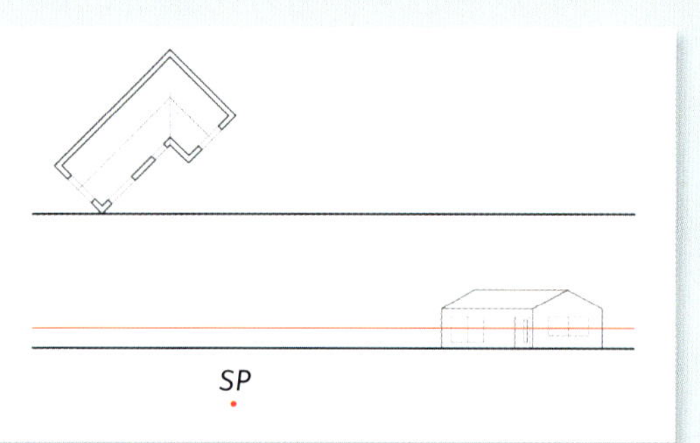

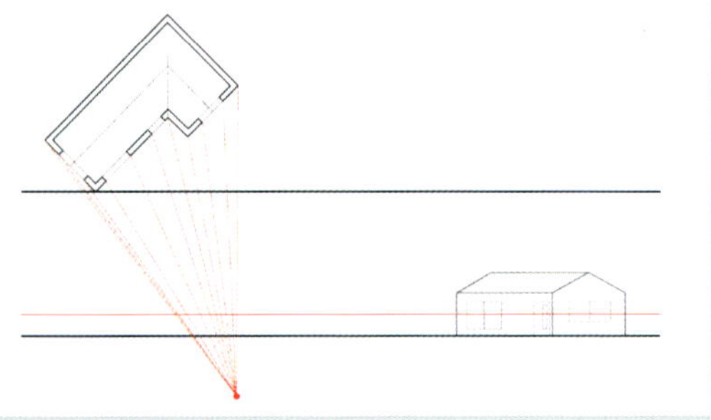

3. Refer to the plan and project faint construction lines from both sides of the building facing the picture plane, including the door and windows, back to the station point.

4. This step is key: to locate the vanishing points, use the station point and draw lines parallel to the left and right sides of the plan that just touch the picture plane. Where these lines meet the picture plane, drop verticals down to the eye level. These intersections establish your vanishing points.

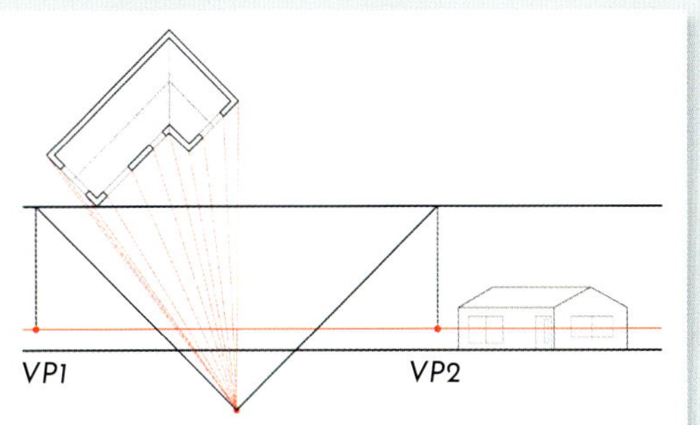

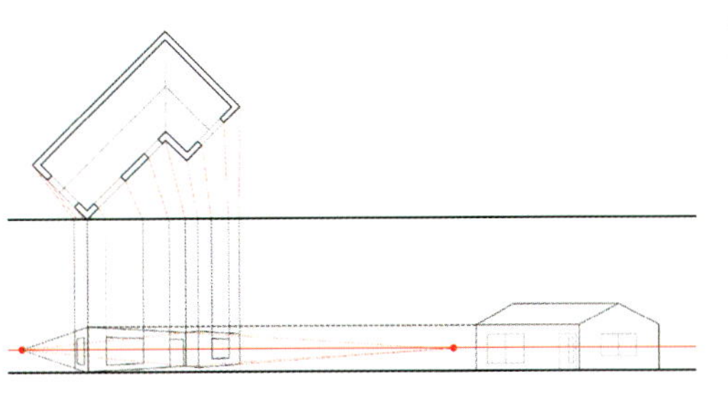

5. Project into perspective: from the plan corner touching the picture plane, drop a vertical line to the ground line. Then project a horizontal line from the top of the wall on the elevation (excluding the pitched roof) until it meets this leading vertical line. This establishes the leading corner of the building in two-point perspective. Extend lines from each corner of the plan to meet the picture plane, then drop verticals to place corners, windows, and the door. From here, use the vanishing points to complete the rectangular form, without the roof, in two-point perspective.

6. Finally, extend the roof line from the elevation to the vertical leading edge, then project it back to the left-hand vanishing point. Extend the left-hand wall until it meets this receding line – this creates the perspective framework for the pitched roof. By drawing diagonals across the gable end wall, you can locate its centre point and position the roof pitch correctly in perspective. Use diagonals again to find the apex of the opposite gable elevation. The angle where the two roof planes meet is determined by the perspective, giving the correct roof slope automatically.

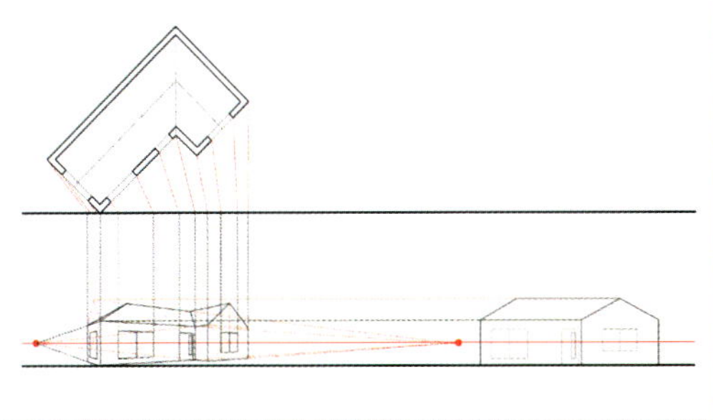

THE FUTURE OF PERSPECTIVE DRAWING

While digital tools have transformed architectural illustration, freehand perspective drawing continues to hold a vital place in the creative process. Its immediacy and expressive quality bring a character and charm that software alone can't replicate.

Historically, architectural drawings were created entirely by hand, and although most practices now rely on digital methods, some still value freehand drawings to develop ideas quickly, for creativity and human quality. Looking ahead, the most compelling illustrations will likely come from combining digital accuracy with the clarity and authenticity of hand-drawing.

Software such as Adobe Photoshop and apps including Procreate and Concepts, make it easy to adjust, colour, and present work in ways that hand-drawing alone can't match. Used together, freehand and digital methods complement each other and offer a more versatile approach to architectural illustration.

A wide range of digital painting and rendering platforms are now available, each with unique advantages for designers and illustrators. It is worth exploring programmes such as Blender, Affinity Designer, and Corel Painter, for sketching, colouring, and layering work in ways that complement traditional drawing. Alongside these, architectural and spatial design programmes such as AutoCAD, SketchUp, and Revit have expanded the role of drawing; applying traditional principles like perspective to support 3D modelling, and realistic visualisations.

It's also worth recognizing (though it falls outside the scope of this book) that perspective drawing plays an important role in other fields, particularly games design. Many games rely on perspective techniques to create convincing environments, and the rise of virtual reality (VR) has pushed this even further. With VR, artists can work in an immersive 3D canvas, using a headset to draw in space rather than on a flat page. Games developers are already combining traditional perspective principles with these new technologies, and this is only the beginning. As artificial intelligence (AI) platforms such as Midjourney – software that generates images from text prompts – continue to evolve, they will no doubt add new possibilities to the future of drawing.

One of the biggest advantages of new drawing technology can be seen in architecture itself. The fluid and complex parametric forms developed by Zaha Hadid Architects are a well-known example of what advanced 3D tools make possible – shapes that would be extremely difficult to imagine, let alone build, using traditional hand-drawn methods.

These digital tools expand what hand-drawn illustration can do, making it easier to combine the immediacy and expressiveness of freehand work with the accuracy and flexibility of digital rendering. Even so, nothing replaces the satisfaction of being fully engaged in drawing, when a sketch comes together naturally on the page. Whatever direction digital illustration takes, I believe there will always be value in understanding – and applying – the basic principles of perspective drawing.

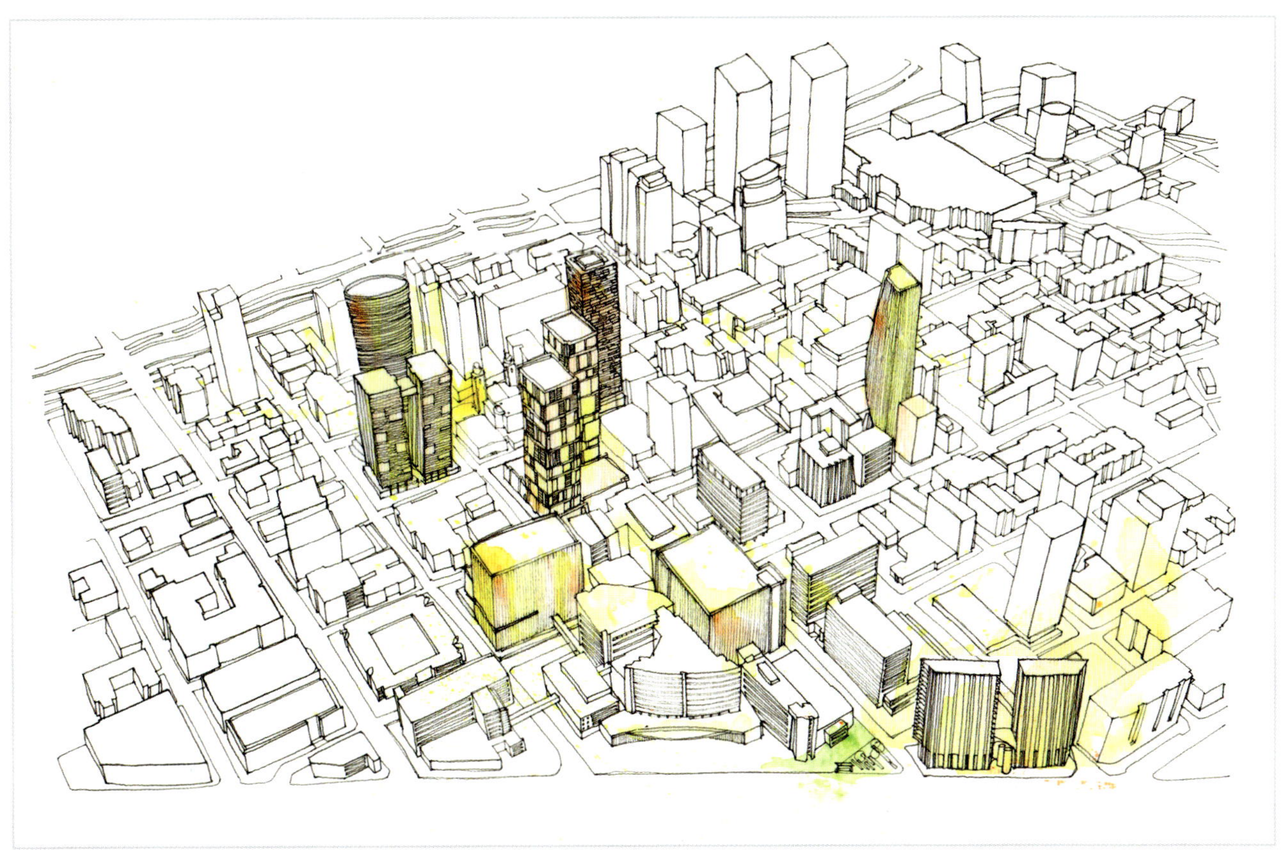

Architectural illustration showing a bird's-eye view of the First Hill neighbourhood masterplan in Seattle, USA, commissioned by Canadian real-estate developer Westbank Corp.

GLOSSARY

Architrave: the decorative frame surrounding a door, window, or opening.

Atmospheric perspective: when things further away appear lighter and less detailed – faded even.

Bird's-eye perspective: a viewpoint from above, looking down on a scene, making objects appear smaller.

Ceiling line: the height of the ceiling.

Cone of vision: the recommended viewing area in perspective drawing, typically about 60 degrees wide, within which objects appear natural and undistorted. Outside this area, distortion increases.

Curvilinear perspective: uses curved lines instead of straight lines to show a wide, panoramic view (often referred to as a fish-eye lens view).

Elevation: a flat straight-on view of one side of a building.

Eye level: a notional horizontal line which represents the viewer's eye level. It is important to note that eye level is unique to each viewer. In one- and two-point perspective the vanishing point or points always sit on this line.

Field of view: the total area visible to the viewer. In perspective drawing, the cone of vision represents the central part of this area.

Foreshortening: when objects appear compressed because they are angled towards or away from the viewer.

Furnitecture: where furniture meets architecture! Interior design terminology to describe built-in or structural furniture that forms part of the architectural design of the space.

Ground line: the horizontal line where a wall or similar object meets the ground.

Ground plane: is the physical surface where objects rest.

Horizon line: see *Eye level.*

Line of sight: the direction we are looking.

Linear perspective: uses straight converging lines and vanishing points to create the illusion of depth on a flat surface.

Loggia: from the Italian word for "lodge"; simply a covered outdoor walkway or gallery, open on one side to the elements. You'll often see them running along the front of a building, though sometimes they stand alone as a feature. The open side is usually held up by columns or arches, which often makes them decorative and fun to draw.

Measuring point: an additional vanishing point used to accurately measure depth and spacing in perspective drawings.

Multi-point perspective: a type of perspective that uses more than two vanishing points to show objects from complex or multiple angles.

Natural perspective: describes the way we experience depth and distance without applying the formal rules of perspective drawing. In art, natural perspective means creating a sense of space and depth that feels convincing and true to how we actually see the world.

One-point perspective: uses one vanishing point and is good for drawing straight-on views.

Picture plane: the imaginary vertical plane which represents your drawing surface; think of it like looking through a window.

Plan: a technical drawing drawn from above showing walls, doors, and furniture.

Plan view: top-down drawing of a building used to construct a perspective drawing.

Rule of thirds: a principle widely used in both art and photography that involves using a grid of nine equal sections by two vertical and two horizontal lines. The idea is to position your main subject along one of these lines – or ideally at one of the four intersection points – rather than in the very centre.

Sectional perspective: where a building is cut through to show the interior in three-dimensional view using the principles of one-point perspective.

Springing line: the starting point of a curve in an arch.

Station point: the exact point from which the viewer observes the scene.

Three-point perspective: uses three vanishing points and is good for extreme high or low views (for example looking up at skyscrapers or down from above).

Two-point perspective: uses two vanishing points and is good for drawing corners of buildings.

Trapezoidal: describes a shape similar to a trapezoid – a four-sided figure with at least one pair of parallel sides. It usually refers to the form or geometry of a building element or space.

Vanishing point: the point on the eye level or horizon line, where parallel lines appear to meet in the distance.

Wireframe effect: a drawing style that shows only the structural edges of a form in perspective, without surfaces or shading, so the drawing appears like a transparent structure made from thin wire.

Worm's-eye perspective: a viewpoint from below, looking up at a scene, making objects appear taller.

ABOUT THE AUTHOR

Simone Ridyard (@simoneridyard) is an internationally recognized architectural illustrator and senior lecturer in Interior Design at Manchester Metropolitan University. Known for her work with Urban Sketchers, she is the author of *Archisketcher: Drawing Buildings, Cities and Landscapes* (North Light Books, 2015), and has contributed to numerous publications on urban sketching. She teaches perspective drawing workshops across the UK and internationally – and has collaborated on high-profile art campaigns with The Big Draw (UK), Winsor & Newton, and Derwent.

Phil Griffin is a Manchester-based freelance writer, broadcaster, and curator with special interest in architecture and urban issues.

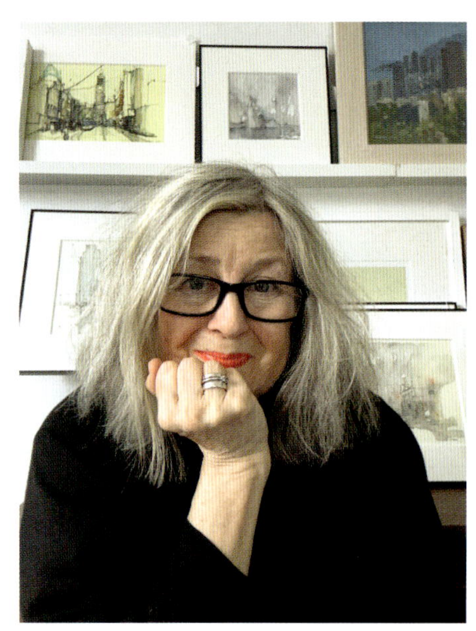

CONTRIBUTORS

The author and the publisher would like to thank the following artists who have allowed their artwork to be featured in this book:

Carlos Almeida, Sketchviews Studio (@sketchviews): 57, 118, 125 (also text on 124)

Darman Angir (Darman Sketcher) (@darman_sketcher): 50–51, 88 (also text on 50)

Lois Blackwell (@loblackwell.design): photo on 76

Stephanie Bower, Architectural Illustration (@stephanieabower): 30, 61, 106–107, 126 (also text on 106)

Phil Dean, Shoreditch Sketcher (@shoreditchsketcher): 79, 80–81 (also text on 80)

Paul Heaston (@paulheaston): 46, 63, 65 (also text on 64)

Karen Jones, KJ Art Studio (@karenjjonesjohn): 40, 48, 78, 84–85 (also text on 84)

Caroline Smith, Urban Sketchliner (@urban_sketchliner): 14, 27, 60 (also text on 26)

Stephen Travers Art (@stephentraversart): 45, 53, 69, 108 (also text on 44)

Willie Watt, Director at Nicoll Russell Studios (@thearchitectssketchbook): 56, 69, 74–75 (also text on 74)

INDEX

A DAVID AND CHARLES BOOK

David and Charles is an imprint of
David and Charles, Ltd Suite A, Tourism House,
Pynes Hill, Exeter, EX2 5WS

EU GPSR Authorised Representative:
Logos Europe, 9 rue Nicolas Poussin,
17000, La Rochelle, France
Email: Contact@logoseurope.eu

First published in the UK and USA in 2026

A catalogue record for this book is available from the British Library.

ISBN-13: 9781446316573 hardback
ISBN-13: 9781446316580 EPUB

This book has been printed on paper from approved suppliers and made from pulp from sustainable sources.

Printed in China through Asia Pacific Offset for:
David and Charles, Ltd
Suite A, Tourism House, Pynes Hill, Exeter, EX2 5WS

10 9 8 7 6 5 4 3 2 1

Publishing Director: Ame Verso
Senior Commissioning Editor: Nigel Browning
Publishing Manager: Jeni Chown
Editor: Victoria Allen
Copy Editor: Melanie Robinson
Lead Designer: Sam Staddon
Designer: Kirsty Kaye
Pre-press Designer: Susan Reansbury
Production Manager: Beverley Richardson

David and Charles publishes high-quality books on a wide range of subjects. For more information visit www.davidandcharles.com.

Share your art with us on social media using #dandcbooks and follow us on Facebook and Instagram by searching for @dandcbooks.

Layout of the digital edition of this book may vary depending on reader hardware and display settings.